Anglo-Saxon Studies 18

COGNITIVE APPROACHES TO OLD ENGLISH POETRY

Anglo-Saxon Studies

ISSN 1475–2468

'Anglo-Saxon Studies' aims to provide a forum for the best scholarship on the Anglo-Saxon peoples in the period from the end of Roman Britain to the Norman Conquest, including comparative studies involving adjacent populations and periods; both new research and major re-assessments of central topics are welcomed.

Books in the series may be based in any one of the principal disciplines of archaeology, art history, history, language and literature, and inter- or multi-disciplinary studies are encouraged.

Proposals or enquiries may be sent directly to the editors or the publisher at the addresses given below; all submissions will receive prompt and informed consideration.

Professor John Hines, Cardiff School of History and Archaeology, Cardiff University, Colum Drive, Cardiff, Wales, CF10 3EU

Dr Catherine Cubitt, Centre for Medieval Studies, University of York, The King's Manor, York, England, YO1 7EP

Boydell & Brewer, PO Box 9, Woodbridge, Suffolk, England, IP12 3DF

Previously published volumes in the series are listed at the back of this book

COGNITIVE APPROACHES TO OLD ENGLISH POETRY

Antonina Harbus

D. S. BREWER

First published 2012
D. S. Brewer, Cambridge

ISBN 978-1-84384-325-2

D. S. Brewer is an imprint of Boydell & Brewer Ltd
PO Box 9, Woodbridge, Suffolk IP12 3DF, UK
and of Boydell & Brewer Inc.
668 Mt Hope Avenue, Rochester, NY 14620, USA

website: www.boydellandbrewer.com

A CIP catalogue record of this publication is available from the British Library

Typeset by Word and Page, Chester

Printed in Great Britain by
CPI Group (UK) Ltd, Croydon CR0 4YY

Contents

Figures

Preface and Acknowledgements

Cognitive approaches to literature offer new and exciting ways of interpreting texts and mentalities, by bringing ideas and methodologies from Cognitive Science into the analysis of literature and culture. This genuinely multidisciplinary group of approaches is of particular value in relation to understanding the texts of remote societies, but has to date made very little impact on Anglo-Saxon Studies. This book maps out this new field, explains its relevance to Old English Literary Studies, and demonstrates in practice its application to a range of key vernacular poetic texts.

This book adapts some key ideas from three related fields – Cognitive Literary/Cultural Studies, Cognitive Poetics and Conceptual Metaphor Theory – and applies those ideas to Old English literature, in concert with more familiar models derived from Literary Analysis, Stylistics and Historical Linguistics. The new frameworks for meaning-making arising from this combination allow several new ways of thinking about Old English literature to emerge. They permit systematic means of examining and accounting for the conceptual structures that underpin Anglo-Saxon poetics, as well as fuller explorations, at the level of mental processing, of the workings of literary language in context. The result is a set of approaches to interpreting Anglo-Saxon textuality through detailed studies of the concepts, mental schemas and associative logic implied in and triggered by the evocative language and meaning structures of surviving texts.

This study traces the arc of imaginative connections through Old English texts, and makes explicit the shared cultural models embedded in conceptual metaphors and other textual features, thereby revealing the underlying templates for understanding human experience, sensory phenomena and mental processes that prevailed in Anglo-Saxon literate cultures. This multi-pronged approach has the related benefit of also accounting for cross-cultural intelligibility, by exploring the psychological hard-wiring and common embodied experience that allows us to understand those same texts today. This disclosure of Anglo-Saxon mentalities is immediately useful for understanding textual rhetoric and the dynamics of meaning, but also more broadly illuminates literary practices and provides a fuller picture of the didactic properties and social functions of poetry across a huge diachronic sweep.

This book came into being as a result of several fruitful discussions I engaged in at the 2009 meeting of the International Society of Anglo-Saxonists, held at St John's, Newfoundland, where I presented a talk from which Chapter 2 below has been developed. I am grateful to the host of that congress,

William Schipper, and its focused attendees, for creating such a congenial context in which inspirational academic exchanges were possible, and where I was able to see how to bring together several lines of my ongoing research inquiries. Particular thanks must go to John Hines for encouraging me at that congress to pursue this project for this series, and also to Toni Healey for her warm encouragement of my inquiries into cognitive studies of Old English literature.

It is likewise a pleasure to thank the other people who have supported the completion of this project and who have assisted me in that task: in particular, my colleagues at Macquarie University, both in the English Department and beyond, especially John Sutton at the Macquarie Centre for Cognitive Science for inviting me to colloquia and embracing the idea of exchanges between early literature and scientific inquiries into the mind. I am fortunate to have had the financial assistance of The Australian Research Council in the production of this book, for which I am grateful, and the research assistance such support enabled, carried out excellently and graciously by Michelle Hamadache, who assisted in the formatting and checking of the footnotes and bibliography, and with proofreading. My thanks too go to the publications wherein portions of this book have appeared previously: a preliminary version of portions of Chapter 1 was published in *Parergon* 27.1 (2010), 13–26, and parts of Chapter 2 in *Anglo-Saxon England* 39 (2011), 21–42.

Finally, I am deeply appreciative of the ongoing personal support, mind-reading and thought-provoking professional critique of Andrew Gillett, and the artistic skills, generosity and perceptiveness of Dora Gillett, who created the cover image, and helped me to navigate my way through Psychology.

Abbreviations

BT, BTS	*An Anglo-Saxon Dictionary*, J. Bosworth and T. N. Toller (Oxford, 1989); *Supplement*, T. N. Toller (Oxford, 1921); *Enlarged Addenda and Corrigenda*, A. Campbell (Oxford, 1972).
DOE	*The Dictionary of Old English*, ed. Antonette diPaolo Healey: http://www.doe.utoronto.ca/index.html

For all poetic texts, other than *Beowulf*, *The Anglo-Saxon Poetic Records*, ed. G. P. Krapp and E. van K. Dobbie (New York, 1931–42), is used. For *Beowulf*, the edition used is: R. D. Fulk, Robert E. Bjork and John Niles, eds, *Klaeber's Beowulf*, 4th edn (Toronto, 2008). For prose texts, the standard editions are used and full details are provided in the notes. Unless stated otherwise, all modern English translations are my own.

1

Introduction

Hwæt! ic swefna cyst secgan wylle,
hwæt me gemætte to midre nihte,
syðþan reordberend reste wunedon.
. . .
ic þær licgende lange hwile
beheold hreocearig hælendes treow,
oððæt ic gehyrde þæt hit hleoðrode.

(*The Dream of the Rood* 1–3, 24–6)

Listen! I will recount the choicest of dreams that came to me in the middle of the night, after speech-bearers had gone to bed . . . . Lying there for a long while, I beheld the splendid tree of glory, until I heard that it spoke.

The vernacular writer of *The Dream of the Rood* conceptualises the human mind, and its ability to apprehend textual representations of dream images and imaginary worlds of scintillating animated crosses, in distinctive ways that are at least partially available to us today. Without some access, not only to those concepts, but also to common human mental functions for processing the text, we would not be able to make sense of these lines, nor have an emotional, intellectual, or aesthetic reaction to them. We would not be able to appreciate and synthesise, for example, the Christian cultural references, the generic dream vision features, or the poetic counterpoint of silent people ('reordberend', speech-bearers) and a speaking cross, let alone the conceptual metaphor of time as space, whereby a night can have a middle. Our very ability to create a coherent response to the text relies on our appreciation of both culturally and individually specific aspects of the inner life embedded in the text, but also requires us to deploy universal human cognitive functions that have changed very little, if any, in the thousand-odd years since the poem was written down in the Vercelli Book.[1] Because the creation and reading of literary texts are cognitive acts requiring multiple processing, every literary encounter is determined by common human mental capacities enacted by an individual mind in specific sociocultural contexts, and so relies on the interaction of literature, culture and cognition.

1 Dated by most scholars to the late tenth century. The relationship of this poem to lines on the Ruthwell Cross, and the wider phenomenon of inscribed outdoor preaching crosses in Anglo-Saxon England, provides a further level of complexity to the multiple speaking positions and thematic interest in utterance in this poem. Similarly, the interest in articulate objects recounting their histories is manifest in many of the Exeter Book riddles. These matters are discussed further in Chapter 3.

The exploration of how the multitasking mind works during that literary encounter is a chief concern of cognitive approaches to literature. These approaches lend themselves particularly to the exploration of texts from remote cultures, but scholars of Anglo-Saxon literature and culture have barely begun to explore vernacular psychology or to deploy the concepts and analytical methods offered by emerging cognitive frameworks for interpreting literary texts. Anglo-Saxonists have only relatively recently made sustained inquiries into vernacular mentalities – including ideas about the mind and the self, how these two entities interact in the individual person – and have barely started considering the implications of various mind schemas for interpreting texts produced in this culture. This line of inquiry was introduced by Peter Clemoes's study of the similarity between patristic and vernacular mind motifs in *The Wanderer* and *The Seafarer*, but was not immediately developed by others.[2] Malcolm Godden's article, 'Anglo-Saxons on the Mind', brings philosophy of mind into the discussion, by arguing that the poets of these two texts recognise two separate 'centres of consciousness', the mind and the self, but, again, it has not been exploited fully.[3] A few later studies have investigated this distinction further, contributing to recent debate on the culturally specific idea of the self in the fields of the Social Sciences[4] by bringing Anglo-Saxon evidence into the discussion.[5] While many studies of the mind and the self have been lexical or at least linguistic in origin,[6] and others have approached medieval mentalities via poetic language

2 Peter Clemoes,'*Mens absentia cogitans* in *The Seafarer* and *The Wanderer*', in *Medieval Literature and Civilization: Studies in Memory of G. N. Garmonsway*, ed. D. A. Pearsall and R. A. Waldron (London, 1969), pp. 62–77. Clemoes develops his earlier line of inquiry in *Interactions of Thought and Language in Old English Poetry* (Cambridge, 1995).

3 Malcolm Godden, 'Anglo-Saxons on the Mind', in *Learning and Literature in Anglo-Saxon England*, ed. Michael Lapidge and Helmut Gneuss (Cambridge, 1985), pp. 271–98. See Chapter 6 below.

4 On the culturally specific self discussed without specific reference to Anglo-Saxon England, see Richard A. Shweder and Edmund J. Bourne, 'Does the Concept of the Person Vary Cross-Culturally?', in *Culture Theory: Essays on Mind, Self, and Emotion*, ed. Richard A. Shweder and Robert LeVine (New York, 1984), pp. 158–99. For the Spanish/English comparison, see José M. Martín Morillas, 'The Concept of Self: Some Cognitive-Cultural Considerations concerning its Categorization and Expression in Spanish and English', *Language Design* 2 (1999), 1–21.

5 Eugene Green, 'Speech Acts and the Question of Self in Alfred's *Soliloquies*', in *Interdigitations: Essays for Irmengard Rauch*, ed. Gerald F. Carr, Wayne Harbert and Lihua Zhang (Frankfurt am Main, Bern, and New York, 1998), pp. 211–18; Holly Elizabeth Jagger, 'Body, Text and Self in Old English Verse: A Study of "Beowulfian" and "Cynewulfian" Rhetoric', Ph.D. thesis, University of Toronto, 2002, DAI 63A (2003); Ronald J. Ganze, 'Conceptions of the Self in Augustine, King Alfred and Anglo-Saxon England', Ph.D. thesis, University of Oregon, 2004, DAI A65/11 (2005); and Antonina Harbus, 'The Medieval Concept of the Self in Anglo-Saxon England', *Self and Identity* 1 (2002), 77–97; and 'Thinking in Metaphors: Figurative Language and Ideas on the Mind', *Sydney Studies in English* 30 (2004), 3–20.

6 Michael Joseph Phillips, 'Heart, Mind, and Soul in Old English: A Semantic Study', Ph.D. thesis, University of Illinois, 1986, DAI 46A (1986); Michiko Ogura, '*Him self, him selfe,* and *him selfa*: A Reflexive Pronoun + Uninflected or Nominative Self', *Studia Neophilologica* 60 (1988), 149–57; Simon Nicholson, 'The Expression of Emotional Distress in Old English Prose and Verse', *Culture, Medicine and Psychiatry* 19 (1995), 327–38; Soon-Ai Low, 'Approaches to the Old English Vocabulary for "Mind"', *Studia Neophilologica* 73 (2001), 11–22; Thomas W. Stewart Jr, 'The Mind and Spirit of Old English *mod* and *fer(h)ð*: The Interaction of Metrics and Compounding', *Ohio State University Working Papers in Linguistics* 52 (1999), 51–62; and

and style,[7] or spirituality,[8] several studies more explicitly incorporate a range of ideas drawn from disciplines such as Philosophy and Psychology, without necessarily engaging with methodologies and data from those fields. For example, Michael Lapidge has demonstrated the *Beowulf*-poet's interest in the workings of the human mind; Soon-Ai Low has shown the psychological focus underpinning *Guthlac B*; and I have argued more broadly for a cultural focus on the mind evident in Anglo-Saxon poetic texts.[9] Leslie Lockett's recent monograph, which traces the evolution and character of Anglo-Saxon ideas about the mind, has taken the debate further, arguing for a 'hydraulic' model of the mind that is essentially realist rather than metaphoric,[10] and makes some use of ideas from 'cognitive linguistics and transcultural psychiatry',[11] but stops short of a cognitive approach to the subject.

These and other studies demonstrate that considerations of pre-modern mentalities can fruitfully draw upon fields additional to literary and linguistic ones. They can canvas the complex subject of mentalities in a more multifaceted way than more traditionally historical studies.[12] Indeed, the nature of the topic insists more than most on the importation and deployment of other disciplinary ideas and methodologies, a cross-disciplinary potential with wide-ranging implications. For example, philosophical and psychological studies of the mind and the self, both medieval and modern, have proved particularly useful for shaping and informing debate on how Anglo-Saxon writers conceived the mind, thereby bringing to light the implications of those ideas for literary and socio-cultural interpretation. Similarly, the growing imperative to historicise literary production and analysis insists on the mutual consideration of cultural and historical context and the literary text. More specifically, the impact of cultural ideas about the mind on textual production has been established as a dynamic worthy of closer investigation, itself

Rūta Šileikytė, '*In Search of the Inner Mind*: Old English *Gescead* and Other Lexemes for Human Cognition in King Alfred's *Boethius*', *Kalbotyra* 54.3 (2004), 94–102.

7 David Wayne Morse, 'A Natural Progression: Cognitive Metaphor as a Structuring Principle in Old English Poetry', Ph.D. thesis, University of Southern California, 1994, DAI 55A (1994); Michael Eugene Matto, 'Containing Minds: Mind, Metaphor, and Cognition in Old English Literature', Ph.D. thesis, New York University, 1999, DAI 59A (1999); and John C. Ford, 'A New Conception of Poetic Formulae Based on Prototype Theory and the Mental Template', *Neuphilologische Mitteilungen* 103 (2002), 205–26.

8 Scott DeGregorio, 'Texts, *Topoi* and the Self: A Reading of Alfredian Spirituality', *Early Medieval Europe* 13 (2005), 79–96. For the later period, see the articles in *Image and Imagination of the Religious Self in Late Medieval and Early Modern Europe*, ed. R. Falkenberg, W. S. Melion and T. M. Richardson (Turnhout, 2003).

9 Michael Lapidge, '*Beowulf* and the Psychology of Terror', in *Heroic Poetry in the Anglo-Saxon Period: Studies in Honour of Jess B. Bessinger Jr*, ed. Helen Damico and John Leyerle, Studies in Medieval Culture XXXII (Kalamazoo, 1993), pp. 373–402; Soon-Ai Low, 'Mental Cultivation in *Guthlac B*', *Neophilologus* 81 (1997), 625–36; Antonina Harbus, *The Life of the Mind in Old English Poetry*, Costerus n.s. 143 (Amsterdam and New York, 2002).

10 Leslie Lockett, *Anglo-Saxon Psychologies in the Vernacular and Latin Traditions* (Toronto, 2011). An alternative perspective to Lockett's is provided in Chapter 2 below.

11 Ibid., p. 9

12 Basil Clarke, *Mental Disorder in Earlier Britain: Exploratory Studies* (Cardiff, 1975); and Charles M. Radding, 'Evolution of Medieval Mentalities: A Cognitive-Structural Approach', *American Historical Review* 83 (1978), 577–97.

drawing upon questions of long pedigree. Historians of mentalities such as Marc Bloch and Michel Foucault have long since established the value of investigating the cultural structures that determine mental activity, in order to demonstrate how closely related the two are and how they fluctuate together.[13] A similar argument for the culture-cognition nexus has been made from the scientific perspective by Michael Tomasello, who demonstrates how cognitive functions are indebted specifically to cultural developments.[14]

Cognitive approaches

In response to these transdisciplinary developments within literary and linguistic studies, and a developing interest in the psychological logistics of textual meaning-making, a new and vigorous group of research fields has arisen in the academy, bringing concepts and methods from Cognitive Science into the service of textual analysis. This still-developing set of fields signals its core interest in mental processing through the use of the definitive term 'cognitive' in labels. The two strands may be grouped under the linguistic – 'Cognitive Poetics' – and the literary – 'Cognitive Literary Studies' (more recently known as 'Cognitive Cultural Studies'), though there are many different interpretive frameworks and analytical practices deployed under these rubrics and their territorial boundaries overlap, particularly in relation to the study of metaphor. Together, they comprise a cognitive turn in textual studies in which many scholars are now involved, in several different and mainly separate approaches to textual and linguistic meaning.[15]

Cognitive approaches offer a new group of perspectives to literary interpretation, and are especially invigorating in any attempt to understand texts produced in remote cultures, such as those written in Anglo-Saxon England. All cognitive approaches to literary criticism focus on the mental processes at work during meaning-making, especially the logistics of interpretation in the mind of the recipient, but for a comprehensive blended method to be of use to literary study, it should also consider the historically situated processes undertaken by the producer of the text. Such an approach would ideally deploy Cognitive Science in order to explore the relationship between mind and meaning from the perspective of the mind and its products as being the

13 See Patrick Hutton, 'The History of Mentalities: The New Map of Cultural History', *History and Theory* 20 (1981), 237–59. Hutton specifically describes this branch of enquiry 'not as a history of ideas but a history of the mind', at p. 238.

14 Michael Tomasello, *The Cultural Origins of Human Cognition* (Cambridge, MA, and London, 1999).

15 Besides these research areas, the fields of Cognitive Linguistics and Cognitive Rhetoric focus on reading and writing as acts of the human mind but remain constrained by traditional field boundaries. This lack of cross-disciplinary exchange is noted by Monika Fludernik, who points out that 'the field at the moment resembles a group of construction sites . . . the different cognitive approaches show no sign of coalescing' ('Narratology in the Twenty-First Century: The Cognitive Approach to Narrative', *Publications of the Modern Language Association of America* 125.4 [2010], 924–30, at p. 927).

result of the close and dynamic interplay of culture and biology.[16] The goal of the following discussion is to describe and demonstrate the usefulness and mechanics of the most compelling cognitive approaches, and to suggest that ideas and methods from these approaches can be usefully blended with literary analytic tools as a means of addressing the needs of Anglo-Saxonists in the twenty-first century.

'Cognitive Science' is an interdisciplinary study of the mind that embraces portions of established disciplines including Psychology, Philosophy, Linguistics, Physiology and Computer Science.[17] It particularly draws on Cognitive Psychology, the branch of Psychology concerned with the *thinking* function of the brain: the mental processes of cognition (remembering, reasoning, calculating, classifying and deciding), with less emphasis on the areas treated by other branches of Psychology, such as feeling, perceiving and behaving, though of course there is a good deal of overlap. Because Cognitive Science has a broader remit, it embraces all of these categories of Psychology and more, especially given its involvement in the emerging evidence for the dynamic interaction of culture and cognition. The developing awareness of the interaction of environment with biology in the cognitive life has encouraged this deep and wide interdisciplinarity, the extent of which is still emerging.

This resulting nexus of concepts and methodologies of the loosely defined field now known as 'Cognitive Science' particularly lends itself to a reinvigoration of literary studies through the provision of multiple new perspectives on how meaning is created in the mind within the triad of text, context and reader. In turn, Literary Studies has a great deal to offer Cognitive Science on a number of fronts, including the tools and methodologies for detailed and nuanced readings of texts as both the products of and the triggers for mental processes; and highly developed and diverse means of approaching texts as sites where culture and cognition interact. At a more fundamental level, literary scholars are already very familiar with the idea that texts are repositories of culturally specific ideas about the mind and mentalities, a concept that can be developed further to embrace some consideration of the way in which texts operate within and enact those mental schemas.

16 For a useful overview of this field, see Alan Richardson, 'Studies in Literature and Cognition: A Field Map', in *The Work of Fiction: Cognition, Culture and Complexity*, ed. Alan Richardson and Ellen Spolsky (Aldershot and Burlington, 2004), pp. 1–29. A lengthy annotated bibliography by Alan Richardson and Mary Crane appears at https://www2.bc.edu/~richarad/lcb/home.html#bib. See also Mark Turner, *Reading Minds: The Study of English in the Age of Cognitive Science* (Princeton, 1991); and David Herman, ed., *Narrative Theory and the Cognitive Sciences* (Stanford, 2003); and issues of journals devoted to the subject: *Poetics Today* 23.1 (2002) and *The European Journal of English Studies* 9.2 (2005).

17 See the online entry in the *Stanford Encyclopedia of Philosophy* at: http://plato.stanford.edu/entries/cognitive-science/: 'Cognitive science is the interdisciplinary study of mind and intelligence, embracing philosophy, psychology, artificial intelligence, neuroscience, linguistics, and anthropology. Its intellectual origins are in the mid-1950s when researchers in several fields began to develop theories of mind based on complex representations and computational procedures.' Paul Thagard, 'Cognitive Science', *The Stanford Encyclopedia of Philosophy (Summer 2010 Edition)*, ed. Edward N. Zalta, URL = <http://plato.stanford.edu/archives/sum2010/entries/cognitive-science/>.

Given these opportunities, it is worth considering at the outset one avenue of research within Cognitive Science that is particularly relevant to the literary scholar, and that is 'distributed' or 'extended' cognition, where some aspects of human memory are considered to be stored culturally rather than individually.[18] The external systems that have an impact on highly content-sensitive mental processes include literary texts that are both the product of and a source of influence on the individual mind: 'Embodied human minds extend into a vast and uneven world of things – artefacts, technologies, and institutions – which they have collectively constructed and maintained through cultural and individual history.'[19] In this still developing paradigm, the mind interacts with other minds through artefacts such as literary texts. One way in which that source of influence is felt is in the narrative development of a sense of self or personal identity. In an instance of borrowing from Literary Studies to the Sciences, social and cognitive scientists have deployed concepts and methodologies from narrative theory to the study of the self, including the narrative basis of autobiographical memory and the self.[20]

Following the source of influence in the other direction, the cognitive linguistic and literary fields now emerging adapt ideas and methodologies from Cognitive Science to consider fully and from multiple perspectives the dynamics of text, context and cognition. The ideas are particularly useful in their provision of new tools for considering the distinctive ways in which literary texts provoke imaginative responses in readers: the impact of complex language features on affective and aesthetic processing. In its attention to the reception of rhetoric, the cognitive group of approaches is as deeply embedded in Linguistics as Psychology, and in its earliest manifestations developed out of stylistic analyses.

Language-based approaches: Cognitive Poetics

The linguistic strand of cognitive approaches to literature has grown out of Cognitive Linguistics – where the origin of language use in perception is central[21] – and deploys ideas from Stylistics and Discourse Analysis synthesised with new concepts from Cognitive Science. This line of inquiry is known

18 See Andy Clark, *Supersizing the Mind: Embodiment, Action, and Cognitive Extension* (Oxford, 2011); Philip Robbins and Murat Ayded, eds, *The Cambridge Handbook of Situated Cognition* (Cambridge, 2009). For a specific study of memory, see Amanda J. Barnier, John Sutton, Celia B. Harris, and Robert A. Wilson, 'A Conceptual and Empirical Framework for the Social Distribution of Cognition: The Case of Memory', *Cognitive Systems Research* 9 (2008), 33–51.

19 Barnier *et al.*, 'A Conceptual and Empirical Framework', p. 34

20 See Jerome S. Bruner, *Acts of Meaning* (Cambridge, MA, 1990); K. Nelson, 'Narrative and the Emergence of a Consciousness of Self', in *Narrative and Consciousness: Literature, Psychology and the Brain*, ed. G. D. Fireman, Ted E. McVay, Jr, and Owen J. Flanagan (Oxford, 2003), pp. 17–36; D. Herman, 'Stories as a Tool for Thinking', in *Narrative Theory and the Cognitive Sciences*, ed. Herman, pp. 163–92; and the further discussion of these issues in Chapter 6 below.

21 For brief overviews of this view of language as a reflex of cognition, see Friedrich Ungerer and Hans-Jörg Schmid, *An Introduction to Cognitive Linguistics* (London and New York, 1996), and David Lee, *Cognitive Linguistics: An Introduction* (Oxford, 2001).

as 'Cognitive Stylistics', or more often 'Cognitive Poetics'.[22] The latter was initially Reuven Tsur's phrase for his means of investigating the psychological processing of literariness, in order to account systematically for 'the relationship between the structure of literary texts and their perceived effects'.[23] This approach combines aspects of Stylistics, Structuralism, Discourse Analysis, and even techniques familiar in New Criticism, with a fine-grained analysis of the way information is coded in poetic language, all with a focus on analysing linguistic choices for evidence of cognitive structures and processes. Tsur uses detailed analytical techniques to identify and account for the features of poetic language that activate special 'aesthetic' cognitive processing and exclusively poetic conceptualisations in the mind of the reader. He deploys this data to develop a scientific method for explaining the cognitive and emotional impact of poetry. Tsur thereby claims that his theory can explain, rather than merely categorise, the effects of poetry, unlike (in Tsur's view) George Lakoff's conceptual metaphor theory, discussed below, pp. 25–7.

Tsur's work is enormously valuable, both for presenting highly specific means of categorising and analysing poetic language, and also for galvanising cognitive approaches to literary texts. But one might challenge the fundamental assumption of Tsur's work, that the language of poetry is inherently distinct from other forms of discourse: it may be, but that distinctive quality is claimed rather than demonstrated by Tsur. Prose can be 'lyrical' or rhythmical (in the case of some of Ælfric's texts), there are many forms of *prosimetrum*, and poetry can be 'prosaic', so the genre categories relied on by Tsur can be legitimately problematised by the literary scholar, even if many of his ideas and central tenets are accepted both within Linguistics and beyond.

Tsur's foundational theories have been taken up and widely adapted by a growing group of scholars, including Peter Stockwell, who has produced an accessible and influential introduction to the field, which, together with its partner volume, *Cognitive Poetics in Practice*, lays out the theory and practical application of this new way of thinking about literary meaning.[24] Stockwell defines Cognitive Poetics as a study of 'the process by which intuitive interpretations are formed into expressible meanings',[25] an approach in which he

22 The two terms are synonymous for most scholars: see, for example, Elena Semino and Jonathan Culpeper, 'Foreword', in *Cognitive Stylistics: Language and Cognition in Text Analysis*, ed. Elena Semino and Jonathan Culpeper (Amsterdam and Philadelphia, 2002), pp. ix–xvi, at p. x.

23 Reuven Tsur, *Toward a Theory of Cognitive Poetics*, 2nd edn (Brighton, 2008), p. 1. See also Tsur, *Poetic Rhythm: Structure and Performance* (Berne, 1998). Though Tsur refers to 'literary texts', his focus is more particularly on the literary effect and emotive affect on the reader of poetry.

24 Peter Stockwell, *Cognitive Poetics: An Introduction* (London, 2002); and Joanna Gavins and Gerard Steen, *Cognitive Poetics in Practice* (London, 2003). See also Semino and Culpeper, eds, *Cognitive Stylistics*, which includes an overview article by Reuven Tsur, 'Aspects of Cognitive Poetics', pp. 279–318.

25 Stockwell, *Cognitive Poetics: An Introduction*, p. 8. This book sets out to introduce the theories laid out by many scholars working in the field, with considerable reliance on Tsur's first (1998) edition, *Toward a Theory of Cognitive Poetics*. In his second edition (2008), Tsur emphatically distances his own concepts and interpretive methodology from those of Stockwell, whom Tsur considers not to practice Cognitive Poetics as he understands it, in that Stockwell is

and others deploy the textual analytic tools of Ronald Langacker's 'cognitive grammar', with its 'dual grounding of language in cognition and social interaction'.[26] For Tsur, the cognitive poetic set of theories and practices seeks to explain the processes occurring in the mind during an aesthetic reading of a text, when the literary effect is 'the result of some drastic interference with, or at least delay in, the regular course of cognitive processes'.[27] This idea of delay in, or the disruption of, normal cognitive processing, is a key idea in Cognitive Poetics: it implies that the mind both recognises poetry and deploys, in response to that recognition, specific interpretive strategies that accommodate associative rather than linear logic and synthetic rather than simplistic processing.[28] Cognitive Poetics thereby deploys rhetorical, stylistic and cognitive theories of aesthetic effects, in order to examine the way literary meanings are produced in the mind of the recipient. This analysis is highly complex and can be approached from a number of different perspectives.[29]

For example, Elena Semino's Theory of Mind Styles combines schema, conceptual metaphor and blending theories, to consider 'how language reflects the particular conceptual structures and cognitive habits that characterise an individual's world view'.[30] Catherine Emmott likewise takes a cognitive poetic approach to the analysis of textual meaning to consider the idea of the split self, an idea familiar to the Anglo-Saxonist.[31] A more explicit attempt to bring Linguistics and Literary Studies into closer contact via Cognitive Poetics by Margaret H. Freeman is more closely interested in the creation of literary meaning. One study relies on the idea of 'poetic iconicity', where form and essence merge, in that poetry creates 'sensations, feelings, and images in language that enable the mind to encounter them as phenomenally real'; another more broadly interdisciplinary approach insists upon the value of a cognitive poetic literary analytic approach.[32]

seen at times to rebadge older methodologies as cognitive ones (620). Notwithstanding this perceived misunderstanding and misapplication of Tsur's ideas, it is clear that both scholars have made substantial contributions to the field, and that their disagreement about what a cognitive approach consists of and how it might be enacted demonstrates the flexibility and still developmental qualities of Cognitive Poetics.

26 Ronald W. Langacker, *Cognitive Grammar: A Basic Introduction* (Oxford, 2008), p. vii.

27 Tsur, *Toward a Theory of Cognitive Poetics*, 2nd edn, p. 4.

28 Compare the empirical approach to the creation of meaning adopted in 'psychonarratology', which aims to 'understand the psychological processing of narrative'. See Marisa Bortolussi and Peter Dixon, *Psychonarratology: Foundations for the Empirical Study of Literary Response* (Cambridge, 2002), p. 35.

29 One interesting mode of analysis, not pursued here, is the cognitive approach to Discourse Analysis. One study of particular relevance to extended literary narrative, concerned chiefly with the mental processing of text is Catherine Emmott, *Narrative Comprehension: A Discourse Perspective* (Oxford, 1997).

30 Elena Semino, 'A Cognitive Stylistic Approach to Mind Style in Narrative Fiction', in *Cognitive Stylistics*, ed. Semino and Culpeper, pp. 95–122, at p. 95.

31 Catherine Emmott, '"Split Selves" in Fiction and in Medical "Life Stories": Cognitive Linguistic Theory and Narrative Practice', in *Cognitive Stylistics*, ed. Semino and Culpeper, pp. 153–81.

32 Margaret H. Freeman, 'The Fall of the Wall between Literary Studies and Linguistics: Cognitive Poetics', in *Cognitive Linguistics: Current Applications and Future Perspectives*, ed. Gitte Kristiansen, Michel Achard, Rene Dirven and Francisco J. Ruiz de Mendoza Ibañez (Berlin and New York, 2006), pp. 403–28, at p. 408; and 'Cognitive Linguistic Approaches to Literary Studies: State of the Art in Cognitive Poetics', in *The Oxford Handbook of Cognitive*

Beyond these individual approaches, some of the most dynamic ideas for textual analysis to arise from Cognitive Poetics belong to Text World Theory. This discourse framework, originally developed by Paul Werth, is currently being used by some scholars working in the field of Cognitive Poetics in order to examine the way in which language is processed in the mind to create mental representations of reality systems.[33] This theory – closely related to the Theory of Mental Spaces developed by Gilles Fauconnier,[34] Elena Semino and their collaborators[35] – draws on both Cognitive Linguistics and Cognitive Psychology to consider how a context influences not only a text's production but also its reception: the way in which context-specific text-worlds, or mental models, are produced during human communication.[36] This theory attempts to account for the production and management of these imagined representations during discourse processing, and is a particularly useful tool for discussing the multiple, shifting and blending worlds evoked in poetic discourses and sustained throughout long fictional narratives.

Peter Stockwell takes this theory further in his analysis of the role of emotion in the aesthetics of reading, or 'texture': 'the experienced quality of textuality'.[37] The Text World framework and its deployment in Stockwell's analysis of the 'cognitive basis of both feeling and meaning'[38] are applied in Chapter 4 below in relation to *Wulf and Eadwacer, Genesis B* and *Beowulf* in order to demonstrate its potential for Old English literary interpretation, and its role in the development of less specifically linguistic cognitive approaches to literature.

Cognitive Literary/Cultural Studies

Cognitive approaches to Literary Studies cover a broad range of topics clustered around the possibilities enabled by studying mind and literature together. The field explores how ideas from Cognitive Science might be brought to bear on literary interpretation in order to study literature specifically as the product of, and something that is processed by, the human mind. The most important idea to be borrowed from Cognitive Science is the notion of the

Linguistics, ed. Dirk Geeraerts and Hubert Cuyckens (Oxford, 2007), pp. 1175–1202.

33 Paul Werth, *Text Worlds: Representing Textual Space in Discourse* (London, 1999).

34 Gilles Fauconnier, *Mental Spaces: Aspects of Meaning Construction in Natural Language* (Cambridge, MA 1985; repr. Cambridge, 1994); *Mappings in Thought and Language* (Cambridge, 1997). See also G. Fauconnier and E. Sweetser, *Spaces, Worlds, and Grammar* (Chicago, 1996). This theory is discussed in Chapter 3 below.

35 The analysis of the system of reference and co-reference encoded in specific texts is the central interest of the Theory of Mental Spaces. For an application of these ideas to literary texts, see Elena Semino, *Language and World Creation in Poems and Other Texts* (London, 1997); 'Possible Worlds and Mental Spaces in Ernest Hemingway's "A Very Short Story"', in *Cognitive Poetics in Practice*, ed. Gavins and Steen, pp. 83–98; and 'Blending and Characters' Mental Functioning in Virginia Woolf's "Lappin and Lapinova"', *Language and Literature* 15 (2006), 55–72. See also Semino and Culpeper, eds, *Cognitive Stylistics*.

36 Joanna Gavins, *Text World Theory: An Introduction* (Edinburgh, 2006), p. 8.

37 Peter Stockwell, *Texture: A Cognitive Aesthetics of Reading* (Edinburgh, 2009), p. 1.

38 Ibid. p. 4.

bodily basis of all human experience, including cognitive processing and mental representation. The core concept here, that the thinking mind is located within and operates as part of the human body – embodiment – is combined with the impact of the cultural context on mental processing, to produce a rich and flexible set of approaches to literary understanding. This collection of theories and methodologies pursues issues such as the mental processing of literariness; cognitive perspectives on genre; and the way in which local ideas about the mind inform the creation and reception of literary texts.

These issues, and others, are pursued by scholars working in the field of 'Cognitive Literary Studies' (more recently 'Cognitive Cultural Studies', an MLA Discussion Group under the first name since 1998), whose most prominent members include Ellen Spolsky, Francis Steen, Mark Turner, Lisa Zunshine, Alan Richardson and Patrick Hogan.[39] This development reflects a broader interest in the cognitive-literary alliance, reflected in graduate programs and centres, such as the Purdue University Center for Cognitive Literary Studies,[40] and websites such as *Blending and Conceptual Integration*;[41] *CogWeb: Cognitive Cultural Studies*;[42] and *Literature, Cognition and the Brain*.[43]

Within this rapidly developing field, which considers the interaction of text, culture and biology, a growing body of influential work is emerging, much of which deals with early-modern literature. For example, Ellen Spolsky, F. Elizabeth Hart and Mary Thomas Crane all deploy cognitive approaches to Renaissance texts to demonstrate how constraining culturally situated ways of thinking can be and therefore the extent to which they shape textual production and interpretation.[44] Spolsky's work grows out of her earlier theory that the modularity of the mind leaves gaps in the human cognitive structure, and that highly imaginative literary texts are attempts to bridge those gaps.[45] This idea is a cognitive repositioning of Wolfgang Iser's influential idea of textual gaps, left for the reader to fill and synthesise.[46] Its interesting corollary

39 See *Cognitive Approaches to Literature*, MLA Discussion Group (revised 17 April 2002). Accessed 31 August 2011 at: http://cogweb.ucla.edu/Culture/CAL/index.html.

40 'Drawing on discourses in the sciences and the humanities, the Center promotes and develops research that explores not only an empirical understanding of the mind as found in Psychology, Biology, Neuroscience, and Information Theory, but also its theoretical development in Philosophy, and its representation in Literature, and the discursive relationship between all of the "kinds of minds."' Paula Leverage, Director, *The Center for Cognitive Literary Studies*. Accessed 31 August 2011 at: http://www.cla.purdue.edu/coglit/.

41 Run by Mark Turner, *Blending and Conceptual Integration*, Case Western Reserve University. Accessed 31 August 2011 at: http://markturner.org/blending.html.

42 'A research tool for exploring the relevance of the study of human cognition to communication and cultural studies', Francis F. Steen, *Cogweb: Cognitive Cultural Studies*, UCLA. Accessed 31 August at: http://www.cogweb.org.

43 Alan Richardson, *Literature, Cognition and the Brain*, Boston College, 'featuring research at the intersection of literary studies, cognitive theory, and neuroscience'. Accessed 31 August 2011 at: https://www2.bc.edu/~richarad/lcb/home.html.

44 All articles are in Richardson and Spolsky, eds, *The Work of Fiction*: Ellen Spolsky, 'Women's Work is Chastity: Lucretia, *Cymbeline*, and Cognitive Impenetrability', pp. 51–84; F. Elizabeth Hart, 'Embodied Literature: A Cognitive-Poststructuralist Approach to Genre', pp. 85–106; and Mary Thomas Crane, '"Fair is Foul": *Macbeth* and Binary Logic', pp. 107–26.

45 Ellen Spolsky, *Gaps in Nature: Literary Interpretation and the Modular Mind* (Albany, 1993).

46 For a brief account, see Wolfgang Iser, 'Interaction between Text and Reader', in *The Book History*

is that literature has developed in response to this gap-filling imperative, to rehearse and hone creative interpretive processes that are necessary in other contexts. This theory could account for why relatively strenuous cognitive work is required to make meaning from the evocative language of poetry and highly abstract prose – including the interpretation of image schemas, conceptual blends and figurative language – and might support an argument for the contribution of literary training in the development of critical thinking skills. It also provides some insight into why literature has developed in both oral and scribal cultures particularly to represent as well as to explore human consciousness, while at the same time fulfilling other agendas, such as the transmission of stories and beliefs, as well as more broadly serving didactic or entertainment functions. The matters are currently among those being investigated by a broad range of cognitive literary inquiries, in which biological and cultural evolution are considered in concert.

In attempting to address this issue of the cognitive role of literature, Lisa Zunshine has shown how fruitful a robust interaction between science and literary analysis can be, and like Spolsky, Hart and Crane, she focuses on non-contemporary literature. She demonstrates very convincingly how the particular aptness of the novel for the exploration of human consciousness can be deployed within Cognitive Literary Studies to examine the workings of the human mind and levels of intentionality, and more broadly the way in which literature can be an instrument of cognition.[47] Her work on Theory of Mind, or mind-reading (our ability to predict states of mind from behaviour), shows that this capacity for extrapolating thoughts and feelings from behaviour allows literary texts to exist and to be intelligible to us:

> The very process of making sense of what we read appears to be grounded in our ability to invest the verbal flimsy constructions that we generously call 'characters' with a potential for a variety of thoughts, feelings, and desires, and then to look for the 'cues' that allow us to guess at their feelings and thus to predict their actions.[48]

Zunshine argues that humans are cognitively evolved to make sense of fictional narrative, as well as real-life encounters, through deploying Theory of Mind, and that literary encounters serve a function of fine-tuning this technique of attributing mental states to other people as well as providing the pleasure of '"trying on" mental states *potentially available* to us'.[49] Indeed, one of the chief sources of enjoyment in any literary encounter might be what Alan Palmer describes as 'being told what a variety of fictional people

Reader, ed. David Finkelstein and Alistair McCleery (London and New York, 2002), pp. 291–6.

47 The latter idea – that literary texts can assist in the refinement of cognitive tools for making sense of the world beyond the text – is developed in a similar way in relation to *Beowulf* by one of the few examples of the analysis of Old English textual narrative structures as evidence of narrative thinking: David Herman and Becky Childs, 'Narrative and Cognition in *Beowulf*', *Style* 37 (2003), 177–203.

48 Lisa Zunshine, 'Theory of Mind and Experimental Representations of Fictional Consciousness', *Narrative* 11 (2003), 270–91, at p. 274.

49 Lisa Zunshine, *Why We Read Fiction: Theory of Mind and the Novel* (Columbus, 2006), p. 17.

are thinking',[50] since this is not a facility we have in our everyday interactions with living people. In her exploration of this and other aspects of the representation of the conscious mind in literature, Zunshine raises briefly the phenomenon of diachronic change. The way in which culturally embedded literary texts reflect different responses to the narrative exploitation of our mind-reading ability is summarised as 'how we construct our world both 'culturally' *and* cognitively'.[51] In this view, we read fiction because it 'engages, teases, and pushes to its tentative limits our mind-reading capacity', precisely because it requires us to keep track of the complex levels of intentionality and sources of representations.[52]

Zunshine develops lines of inquiry presented by Mark Turner, who sees Classical Rhetoric and Cognitive Neuroscience as the foundations of Cognitive Literary Studies.[53] Turner seeks to claim that 'the everyday mind is the essentially literary', and that language is not the origin but the product of parable, the consequence of the way our mind is hardwired for story.[54] Turner relies on the idea of conceptual integration, or blending, to explain the mind's capacity for figurative and synthetic thought processes. In this fundamental and widely used cognitive activity, discussed at length in Chapter 4 below, two mental frames or schemas are blended together to make a new meaning. Conceptual blending has become a hugely influential and useful concept in cognitive approaches to literature, with particular relevance to metaphor, whose defining characteristic is the blending of two concept domains into a single figurative expression: conceptual integration.

Conceptual metaphor

As for Cognitive Poetics, the study of conceptual metaphor – as a feature of both language and thought – holds special interest for those working in Cognitive Literary Studies, as evidence of the innately creative human mind, and its associative manner of constructing abstract ideas. Several important collections of articles have been devoted exclusively to cognitive approaches to understanding metaphor in particular: *Poetics Today* 13.4 (1992), 14.1 (1993) and 20.3 (1999). Here, and in other publications, conceptual metaphor is a field studied by cognitive linguists, those working in Cognitive Poetics, Text World theorists, literary scholars of various stripes, and many others. This cognitive approach is premised on the influential idea that metaphors express deep-seated mental ways of organising thought, put forward by

50 Alan Palmer, *Fictional Minds* (Lincoln and London, 2004), p. 10.
51 Lisa Zunshine, 'Richardson's *Clarissa* and a Theory of Mind', in *The Work of Fiction*, ed. Richardson and Spolsky, pp. 127–46, at p. 142.
52 Zunshine, *Why We Read Fiction*, p. 4.
53 Mark Turner, 'The Cognitive Study of Art, Language, and Literature', *Poetics Today* 23.1 (2002), 9–20, at p. 17.
54 Mark Turner, *The Literary Mind: The Origin of Thought and Language* (New York, 1996), especially p. 7 and p. 168. See also Mark Turner and Giles Fauconnier, 'A Mechanism of Creativity', *Poetics Today* 20.3 (1999), 397–418; and *The Way We Think: Conceptual Blending and the Mind's Hidden Complexities* (New York, 2002).

George Lakoff and Mark Johnson, in *Metaphors We Live By*, and developed in subsequent publications.[55] Many scholars now believe that we use metaphors not only to express ourselves in language, but also to formulate ideas: we think in metaphors.[56] In this schema, the most important idea is that of the categorisation of thoughts into groups expressed through conventional metaphors. For example, the conceptual metaphor THE MIND IS A CONTAINER[57] gives rise to expressions such as 'file that thought away in the back of your mind' and 'dredge up a memory'. Lakoff and Johnson show convincingly that such metaphors are pervasive in everyday speech and play a significant role in organising personal and social experience: through their organisational relationships and conventional use, they condition our thought processes and systematically influence behaviour. That is, people actually think in terms of the mind being a container and thereby conceptualise, organise, and access their memories accordingly.

The rich seam of inquiry opened up by the work of Lakoff and Johnson has been instrumental in galvanising cross-disciplinary studies of metaphor, in particular the application of Cognitive Linguistics to the study of literature by a huge range of scholars.[58] Many have brought cultural and communication theory to the study of mind and metaphor, thereby expanding the range of considerations within our understandings of metaphor use and interpretation.[59] The arguments of Lakoff and Johnson have been extremely influential in the investigations in both Cognitive Literary Studies and Cognitive Poetics, where the study of metaphor is of central importance, especially in the way these figurative expressions encode and perpetuate culturally specific models for the mind, or schemas: templates or conventionally assembled sets of properties.

Conceptual metaphor is discussed in detail in Chapter 2 below – particularly the apparently contradictory metaphors of the mind as a container and a wandering entity in Old English texts – but it is worth noting here that the study of mind schemas encoded in recurrent metaphors in Old English poetry is one area in which the cognitive approach has made some impact on

55 George Lakoff and Mark Johnson, *Metaphors We Live By* (Chicago, 1980). See also George Lakoff and Mark Turner, *More than Cool Reason: A Field Guide to Poetic Metaphor* (Chicago and London, 1989).

56 For example, Raymond W. Gibbs Jr, *The Poetics of Mind: Figurative Thought, Language and Understanding* (Cambridge, 1994).

57 The use of small capitals to indicate conceptual metaphors has become conventional in Cognitive Linguistics and other fields, so is used here for clarity.

58 Zoltán Kövecses, in *Metaphor: A Practical Introduction* (Oxford, 2002), brings together the study of language, culture and the mind in a cognitive-linguistic theory of metaphor, and collections of studies provide multiple perspectives on the cognitive approach to metaphor. See, for example, Raymond W. Gibbs Jr and Gerard Steen, eds, *Metaphor in Cognitive Linguistics* (Amsterdam and Philadelphia, 1999); and Antonio Barcelona, ed., *Metaphor and Metonymy at the Crossroads: A Cognitive Perspective* (Berlin and New York, 2000).

59 Naomi Quinn, for example, argues that 'cultural understanding underlies metaphor use' through the 'organizing role of culture in human thought' in 'The Cultural Basis of Metaphor', in *Beyond Metaphor: The Theory of Tropes in Anthropology*, ed. James W. Fernandez (Stanford, 1991), pp. 56–93, at p. 57. Similarly, the link between the use of figurative language and ideology is the key focus of Jonathan Charteris-Black, *Corpus Approaches to Critical Metaphor Analysis* (Houndmills, 2004).

the field. In particular, the habit of conceptualising the mind as a figurative container for thoughts is becoming well recognised in the field. The use of these metaphors, especially in prose texts translated from Latin models, shows them to have been inspired by inherited Christian metaphors, and adapted for vernacular deployment. Building on the work of Lakoff and Johnson on conceptual metaphors, several scholars have contributed to our understanding of the Anglo-Saxon use of the container metaphor for the mind, and, more broadly, have outlined the models of the mind articulated in vernacular poetry,[60] and epistemological premises on the mind underpinning vernacular translations.[61] In an explicit appropriation and thoughtful development of ideas promoted by Lakoff and Johnson, Christian Kay has considered conceptual metaphors transmitted beyond Old English texts.[62] Now that the *Historical Thesaurus*, on which Kay has been editor, has been published, the opportunity has arisen for even more detailed studies of this type, in which the diachronic dimension to conceptual metaphor development and transfer can be more fully investigated.[63]

The embodied mind

Cognitive approaches to the study of literature consider the way in which texts have the capacity to provoke in the recipient complex psychological acts involving memory, conceptual blending, the deployment and adjustment of narrative schemas, and the creation of imagined worlds – all aspects of meaning making analysed within cognitive approaches to literature, and discussed in detail in the chapters which follow. The resulting creation of literary interpretation is a combination of text, culture and cognition, enabled and constrained by the capacity of the human brain. The consideration of all these components is integral to those scholars working within the cognitive project on Evolutionary Psychology or the adapted human mind. The appre-

60 See Antonina Harbus, 'The Maritime Imagination and the Paradoxical Mind in Old English Poetry', *Anglo-Saxon England* 39 (2011), 21–42; Leslie Lockett, 'Corporeality in the Psychology of the Anglo-Saxons', Ph.D. thesis, University of Notre Dame, 2004, DAI-A 65/06 (2004) (and the recent monograph developed from that thesis, cited above n. 10); Britt Mize, 'The Representation of the Mind as an Enclosure in Old English Poetry', *Anglo-Saxon England* 35 (2006), 57–90; and Mize, 'Manipulations of the Mind-as-Container Motif in *Beowulf, Homiletic Fragment II,* and Alfred's *Metrical Epilogue to the Pastoral Care', Journal of English and Germanic Philology* 107 (2008), 25–56. Both Lockett and Mize build on the ideas articulated in Soon-Ai Low, 'The Anglo-Saxon Mind: Metaphor and Common-Sense Psychology in Old English Literature', Ph.D. thesis, University of Toronto, 1998, DAI-A 60/10 (2000); Low, 'Approaches to the Old English Vocabulary for Mind'; Low, 'Mental Cultivation'; and Matto, 'Containing Minds'.

61 For instance, see Miranda Wilcox, 'Alfred's Epistemological Metaphors: *eagan modes* and *scip modes*', *Anglo-Saxon England* 35 (2006), 179–217.

62 Christian J. Kay, 'Metaphors We Have Lived By: Pathways between Old and Modern English', in *Essays on Anglo-Saxon and Related Themes in Memory of Lynne Grundy*, ed. Jane Roberts and Janet Nelson (London, 2000), pp. 273–85.

63 Christian J. Kay, Jane Roberts, Michael Samuels and Irené Wotherspoon, *Historical Thesaurus of the Oxford English Dictionary: With Additional Material from a 'Thesaurus of Old English'* (Oxford and New York, 2009).

ciation of the biological aspect of any mental act has particular relevance for the study of remote cultures. In this schema, mental structures, determined by evolutionary development, 'create a set of species-typical behavioural dispositions'[64] that play out variously in cultural practices and products, with the result that cognitive universals may be manifest in these products, which include literary texts.[65] This idea is central to Cognitive Poetics: Reuven Tsur argues that developments in literary history are determined by deep-seated universals in cognitive functioning, that 'rigid conventions are fossilized cognitive processes'.[66]

One such universal is the physical embodiment of the mind: mental functions and states are the result of brain activity, which is physically situated within, and pervasively influenced by this placement in the human body. The concept of embodiment has been popularised by the widely influential work of Mark Turner, particularly *Reading Minds: The Study of English in the Age of Cognitive Science*.[67] Working under the rubric of 'Cognitive Rhetoric', conceived as an offshoot of Cognitive Linguistics, he argues that the way the mind works arises from the fundamental fact that it exists in a physical body: that embodiment provides an inescapable context of meaning. George Lakoff and Mark Johnson develop a similar argument in *Philosophy in the Flesh: The Embodied Mind and its Challenge to Western Thought*.[68] The study of embodied cognition in the sciences has been pursued most notably by Andy Clark and Antonio Damasio,[69] and recently advanced by many others investigating the formative role of the physical body on human perception.[70] Ray Gibbs has detailed the implications of ideas of an embodied mind for Cognitive Science more broadly, with particular relevance to the study of metaphoric language.[71] In this view, figurative ways of imagining and expressing abstract concepts arise from basic embodied activities such as moving, eating and standing

64 Joseph Carroll, 'Evolutionary Approaches to Literature and Drama', in *Oxford Handbook of Evolutionary Psychology*, ed. R. I. M. Dunbar and Louise Barrett (Oxford, 2007), pp. 637–48. Carroll points out that adaptationist literary critics and cognitive rhetoricians do not always share conceptual and methodological ground (p. 639). The former are often more concerned with the evolved structure of cognitive inclinations, and the latter with brain function and affective responses to literature.

65 See Tony E. Jackson, 'Issues and Problems in the Blending of Cognitive Science, Evolutionary Psychology, and Literary Study', *Poetics Today* 23.1 (2002), 161–79. For an extended consideration of the application of ideas from evolutionary psychology to literary analysis, see Brian Boyd, *On the Origin of Stories: Evolution, Cognition, and Fiction* (Cambridge, MA, and London, 2009).

66 Reuven Tsur, 'Some Cognitive Foundations of "Cultural Programs"', *Poetics Today* 23.1 (2002), 64–89, at p. 84.

67 Turner, *Reading Minds*.

68 George Lakoff and Mark Johnson, *Philosophy in the Flesh: The Embodied Mind and its Challenge to Western Thought* (New York, 1999).

69 Andy Clark, *Being There: Putting Brain, Body, and World Together Again* (Cambridge, MA, 1997); Antonio R. Damasio, *Descartes' Error: Emotion, Reason, and the Human Brain* (New York, 1994); and Damasio's *The Feeling of What Happens: Body, Emotion and the Making of Consciousness* (London, 1999). Damasio's work is particularly compelling in its argument for the bodily role in the emotional life. See also F. J. Varela, E. Thompson and E. Rosch, *The Embodied Mind: Cognitive Science and Human Experience* (Cambridge, MA, 1991).

70 For instance, Shaun Gallagher, *How the Body Shapes the Mind* (Oxford, 2005), and Robert Hanna and Michelle Maiese, *Embodied Minds in Action* (Oxford, 2009).

71 Raymond W. Gibbs Jr, *Embodiment and Cognitive Science* (Cambridge, 2006).

upright, in turn producing durable and adaptable 'image schemata' that underpin the production of new expressions. In an even more wide-ranging study, with particular application to literary texts, Edward Slingerland argues for an imperative shift away from postmodern theory in Literary Studies to an approach to culture that integrates the natural and biological sciences – specifically the embodied view of the person – with humanistic inquiry into the conceptual products of that embodied mind.[72]

These ideas can be applied fruitfully to develop our understanding of mind schemas that inform the representations of consciousness in earlier literary texts,[73] and to make cross-cultural and diachronic comparisons of mental models in order to shed light on the impact of culturally specific factors on embodied and semantic universals.[74] Recent research into mental processes in Cultural Psychology suggests that the way the mind and its attributes and functions are understood is culturally specific to a high degree and varies across time and among societies, because cognition is in part determined by social experience.[75] Studies of the histories of ideas about the mind are also shedding light on the way culture and mind schemas interact.[76]

This idea is embraced in a similar way by the cognitive linguist F. Elizabeth Hart, who insists upon the relevance of a theory of 'the material embodiment of the subject in and through language'.[77] By showing language to be 'imaginatively embodied', Cognitive Linguistics can supply 'a new, metaphor centred model of language . . . that situates the subject within its material world both inside and outside the text'.[78] In other words, the reading subject is both culturally and cognitively situated by their immediate context. This useful contribution to Literary Theory and interpretation brings together key ideas from the fields of Linguistics, Philosophy, Cognitive Psychology and Literary and Cultural Theory to provide a compelling new way of understanding how textual meaning can be both coherent and unstable. Hart's later work succinctly describes the value of Cognitive Literary Studies as investigating 'the mind's

72 Edward Slingerland, *What Science Offers the Humanities: Integrating Body and Culture* (Cambridge, 2008).

73 See the work by Zunshine mentioned above, nn. 48, 49, 51.

74 See, for example, Caroline Gevaert, 'The Anger is Heat Question: Detecting Cultural Influence on the Conceptualisation of Anger through Diachronic Corpus Analysis', in *Perspectives on Variation: Sociolinguistic, Historical, Comparative*, ed. N. Delbecque, J. van der Auwera and D. Geeraerts (Berlin and New York, 2005), pp. 195–208; Marie-Odile Junker, 'A Native American View of the "Mind" as Seen in the Lexicon of Cognition in East Cree', *Cognitive Linguistics* 14.2/3 (2003), 167–94; and Richard E. Nisbett, *The Geography of Thought: How Asians and Westerners Think Differently . . . and Why* (New York, 2003), esp. pp. 39–45.

75 See Petra Hauf and Friedrich Försterling, *Making Minds: The Shaping of Human Minds through Social Context* (Amsterdam and Philadelphia, 2007), esp. p. 2: 'individual minds become social minds by learning to share their mental contents with others'. See Shweder and LeVine, eds, *Culture Theory*; R. D'Andrade, 'A Folk Model of the Mind', in *Cultural Models in Language and Thought*, ed. D. Holland and N. Quinn (Cambridge, 1987), pp. 112–48; and R. A. Sharpe, *Making the Human Mind* (London, 1990).

76 Paul S. Macdonald, *History of the Concept of Mind: Speculations about Soul, Mind and Spirit from Homer to Hume* (Aldershot, 2003).

77 F. Elizabeth Hart, 'Cognitive Linguistics: The Experiential Dynamics of Metaphor', *Mosaic* 28.1 (1995), 1–23, at p. 21.

78 Ibid., p. 2.

substantive indebtedness to its bodily, social and cultural contexts', and human cognition as 'a set of highly imaginative – not logical but figural – processes'.[79]

Paul Hernadi looks at the causal relationship between mental functioning and literary texts from the opposite direction, arguing that literature has played a precise role in the development of human nature.[80] His and other articles published in a special issue of *Poetics Today* devoted to cognitive approaches treat these matters with a keener appreciation of historicist principles, bringing together the idea that literature is the product of both a biologically created mind and one that is culturally constructed. The editors have created a section entitled 'Cognitive Historicism: Situating the Literary Mind', in which articles 'address the complex interrelation of evolved neurocognitive structures and contingent cultural environments with an eye to specific examples of cultural change'.[81] Similarly, Mary Crane synthesises ideas from Cognitive Neuroscience, Linguistics and Cultural History to argue that Shakespeare's use of language is a product of both biology and culture.[82] Even more precisely, Evelyn Tribble's study of cognition on the Shakespearean stage has shown how rich the combination of ideas from Cultural History, Cognitive Science and emerging ideas about distributed cognition can be.[83] The project of 'Cognitive Historicism', by Richardson, Steen, Crane, Tribble and others, combines Cognitive Science with new historicist and other postmodern theories to enable new perspectives on early texts that consider the operation of the embodied mind within culturally constructed psychologies and image schemas.

This analysis of literary meaning-making via the tension between embodied universals (biology) and contextual variables (culture) is an exciting and fruitful new departure for Literary Studies, especially for scholars interested in pre-modern literatures. The representation of consciousness in earlier literary texts, for example, is one area where some exceptionally interesting insights have been presented.[84] Scholars are asking questions such as: 'How does the embodied mind show up in earlier texts, and what cultural factors are at work to cause it to show up in the precise way that it does?'[85] Because cognitive

79 F. Elizabeth Hart, 'The Epistemology of Cognitive Literary Studies', *Philosophy and Literature* 25 (2001), 314–34, at p. 315.

80 Paul Hernadi, 'Why is Literature: A Coevolutionary Perspective on Imaginative Worldmaking', *Poetics Today* 23.1 (2002), 21–42.

81 Alan Richardson and Francis F. Steen, 'Literature and the Cognitive Revolution: An Introduction', *Poetics Today* 23.1 (2002), 1–8, at p. 5. The critique of this special issue of *Poetics Today* by Hans Adler and Sabine Gross, 'Adjusting the Frame: Comments on Cognitivism and Literature', published in the following issue, 23.2 (2002), 195–220, was roundly critiqued itself by both Ellen Spolsky in 'Cognitive Literary Historicism: A Response to Adler and Gross', *Poetics Today* 24.2 (2003),161–83, and the editors of the special volume, Alan Richardson and Francis F. Steen in 'Reframing the Adjustment: A Response to Adler and Gross', *Poetics Today* 24.2 (2003), 151–9.

82 Mary Thomas Crane, *Shakespeare's Brain: Reading with Cognitive Theory* (Princeton, 2004).

83 Evelyn Tribble, *Cognition in the Globe: Memory and Attention in Shakespeare's Theatre* (London, 2011). See also Evelyn Tribble and Nicholas Keene, *Cognitive Ecologies and the History of Remembering: Religion, Education, and Memory in Early Modern England* (London, 2011).

84 See, for example, the publications of Lisa Zunshine cited above (nn. 48, 49, 51).

85 Tony E. Jackson, 'Issues and Problems in the Blending of Cognitive Science, Evolutionary

theories are particularly concerned with the historical as well as the cognitive processes underlying the production of meaning, the task of reconciling the historical variability of texts with their cross-cultural intelligibility is of central importance.[86] But those scholars best placed to engage in this kind of historical inquiry have not yet fully embraced the possibilities of cognitive study. As yet, Cognitive Literary Studies as an approach has just started to have an impact on Anglo-Saxon scholarship and Medieval Studies more broadly, which leaves a range of opportunities for new research in the field.

A new approach to the interpretation of Old English literature

Scholars analysing Old English texts might develop cognitive lines of inquiry in several different ways. One possibility would entail embracing the opportunities of 'delineating the models of mental operations that influenced writers working in earlier historical periods',[87] another would consider the interplay of universal cognitive functioning and human behavioural dispositions with the contingencies of cultural variability. Ellen Spolsky has identified the study of diachronic variability as a core potential of Cognitive Literary/Cultural Studies:

> One of the hardest questions indeed turned out to be how it is that literary systems (both of production and interpretation) change through time, leaving us in the situation of being able to read and appreciate *Beowulf*, for example, but with no chance of that three-thousand-line poetic epic being written now.[88]

Spolsky demonstrates the important contribution to be made by the cognitive approach in seeking to understand how systems of literary meaning that rely on culturally determined mental patterns are nevertheless intelligible beyond that immediate context of textual creation and reception. She argues that cultural patterns constrain human brains – that history mediates knowledge – but that generalised mental processes allow intelligibility to be retained despite cultural change.[89] Cognitive Literary/ Cultural Studies provide a new way of viewing the inter-relationship of literary text and socio-historical context. Elsewhere, Spolsky has put this approach into practice, in her analysis of Chaucer's *Nun's Priest's Tale*,[90] but remains one of the few scholars to interpret medieval literature via the cognitive approach.[91] Mark Turner has likewise

Psychology, and Literary Study', *Poetics Today* 23.1 (2002), 161–79, at p. 175.

86 Richardson and Spolsky, 'Preface', in *The Work of Fiction*, pp. viii–ix.

87 Richardson and Steen, 'Literature and the Cognitive Revolution', at p. 6.

88 Spolsky, 'Cognitive Literary Historicism', p. 164.

89 Ibid., p. 166.

90 Ellen Spolsky, 'Why and How to Take the Fruit and Leave the Chaff', *SubStance* 30 (2001), 177–98.

91 For the few linguistic examples, see David L. Cooper, 'Attractor Dynamics in *Beowulf*', in *Linguistic Attractors: The Cognitive Dynamics of Language Acquisition and Change*, Human Cognitive Processing 2 (Amsterdam and Philadelphia, 1999), pp. 206–41; and Ruta Nagucka, 'Cognitive Approaches for the Understanding of Old English Loanwords', in *Text and Language in Medieval English Prose: A Festschrift for Tadao Kubouchi*, ed. Akio Oizumi, Jacek Fisiak and John Scahill, Studies in English Medieval Language and Literature 12 (Frankfurt am Main,

used this approach on *The Dream of the Rood*,[92] discussed in Chapter 3 below, though he does not consider the implied corollary of his own question ('What is the human mind that it can create and understand a text?'[93]): how can a mind understand a text produced over a thousand years ago?

A more sustained application of cognitive theories and methods by Anglo-Saxon specialists would allow the development of a fuller and more precise attempt to understand not only how the minds of Anglo-Saxon peoples worked and how they understood mental states and functions, but also consider the concomitant implications for textual production and the creation of meaning by both their own and later recipients of their texts. These compelling matters are discussed in the chapters below as a contribution to furthering these specifically cognitive-cultural lines of inquiry. Models of how the mind works must influence the way rhetorical structures and literary production more broadly operate, so it is worth asking how mental schemas and poetic interpretation have been conceived and enacted in remote cultures. Similarly, we might ask what sort of mental operations were anticipated by the creators of literary texts, especially those involving a high degree of imaginative and creative work on the part of the recipient, such as poetic texts, and also how we are able to interpret those same texts over one thousand years later. For example, as suggested above, p. 1, in the case of *The Dream of the Rood*, interpretation of the 'vision within the vision' requires of the recipient a radical revision of concepts from the real world, and the blending of remembered dream experiences with known scriptural information and experience of reading dream narratives.

To appreciate the way in which cognitive approaches complement and extend earlier approaches to textual interpretation, it is necessary to synthesise existing frameworks – especially those relating to narrative, genre and intertextuality – with those more closely associated with Cognitive Science dealt with in the chapters below – memory, self-formation, imagination, conceptual metaphor and blending. This interpretive synthesis provides a flexible and multi-pronged means of analysing the textual encounter with delicacy and precision, and provides fresh perspectives from which to consider the affective impact and creation of meaning in the mind of the recipient.

With the potential new lines of inquiry offered by Cognitive Literary/Cultural Studies and Cognitive Poetics, texts become key sources of information with relevance beyond their own immediate literary meaning: they become instantiations of human cognitive processing and conceptual structures. Anglo-Saxon scholarship will undoubtedly be enriched by probing early-medieval theories of mind more fully than has been done, and by analysing individual vernacular texts or groups of texts for what they reveal about the life of the mind below the level of consciousness: how they function as expressions of the mind that can be apprehended by other minds. This

2005), pp. 163–76.

92 Turner, 'The Cognitive Study of Art, Language, and Literature', p. 10; his discussion of *The Dream of the Rood* is on pp. 14–15 of this article.

93 Turner, *Reading Minds*, p. 16.

approach would entail interpreting how cognitive functions are understood, how mental states are represented, and how behaviour is explained in terms of assumed underlying mental states in a particular cultural context. It necessitates asking questions such as: 'What sort of inferences regarding mental functioning are contained in the written texts?'; 'How is meaning determined and constrained by assumptions about the mental state of the text's recipient?'; and 'To what degree are modern readers cognitively and culturally prepared in a way that differs from Anglo-Saxon readers?'

The chapters below attempt to make a start at responding to these complex and now highly relevant questions with reference to a particular cultural ecology. The poetic and prose texts produced in Anglo-Saxon England, especially the opportunities offered by the associative logic of poetry, lend themselves to this type of analysis. For example, the tendency to refer to the mind rather than to the person as the site of emotion and the recipient of fate in Old English poetry[94] suggests not just a distinctly different conscious way of viewing human perception and the outcome of events from our own, but also an unconscious, culturally determined mind schema that is both alien to us and yet sufficiently recognisable for the text to make sense to us. Indeed, the accessibility of any idea encoded in Old English poetry is mediated by the affective life, as demonstrated by recent work on the philosophy of emotions that describes how emotions function in thought processes.[95]

Similarly, by analogy with our own experience of conceptual blending, we are able to identify and comprehend the blends described by Turner of *The Dream of the Rood*, even though we might not use them ourselves in non-literary contexts, or certainly not express them in Old English. The Anglo-Saxon literary corpus, and in particular its fictional representations of consciousness, is readable to us, but only via the process of linguistic and cultural relocation that operates through translation into modern English (either on the page or in the mind of the experienced reader). Nevertheless, a core degree of intelligibility remains, stemming from our temporal proximity to the Anglo-Saxons in terms of evolutionary biology, and because we share the human experience of an embodied mind and, apparently, a hard-wired predisposition for narrative. This is not to discount the impact of literary training and the way our expectations are determined by prior reading experiences and a host of other factors. We are accustomed, for example, to encountering fictionalised consciousnesses and shifting perspectives – even dream sequences and religious visions – in literary texts, and, once experienced in this sort of mental processing, we readily interpret them in the literary context.

The translated or adapted text provides further interesting opportunities for these lines of inquiry, though pursued only very briefly in the current

94 For example, 'min hyge geomor' ,'my mind was sad', in *The Wife's Lament*, line 17b; and 'ne mæg werig mod wyrde wiðstandan', 'the weary mind cannot withstand fate', in *The Wanderer*, line 15.

95 See K. Oatley and P. N. Johnson-Laird, 'Towards a Cognitive Theory of Emotions', *Cognition and Emotion* 1 (1987), 29–50; and Damasio, *The Feeling of What Happens*.

study. Old English texts for which known Latin sources and analogues exist provide further insight into cross-cultural intelligibility and specific ideas and mental states underlying the transformation in the translation process. Culturally variable theories of knowledge, consciousness and the emotions constrain the possibilities available in the translation process, just as the different lexicons and syntactic structure of the host and target languages constrain linguistic possibilities. Old English translated texts offer a wonderfully rich opportunity for exploring the mental processes at work in the interpretation and linguistic refashioning of a text, because they play out the act of meaning-making within cultural transformation. Old English literary translation is under-represented in scholarly endeavour to date; here is an additional reason to engage with this part of the extant corpus via the opportunities offered by cognitive approaches.

Given the competitive realities of modern academia, and the imperatives of contemporary relevance for any course of study, there might be some inclination to anticipate a relatively bleak future for Old English studies.[96] Yet the scale and vigour of both ongoing and new projects defies any such prediction. In addition to a continued commitment to ongoing projects such as *The Dictionary of Old English* and associated publications and editorial undertakings, digital projects making original documents more widely available,[97] and the growing exploration of Anglo-Latin literature, the future livelihood of Old English scholarship lies in developing fruitful and exciting new avenues of inquiry, such as those offered and just beginning to be explored via Cognitive Literary/Cultural Studies and Cognitive Poetics.[98] By continuing to exploit new transdisciplinary opportunities such as these, Anglo-Saxon Studies can remain relevant, interesting and viable in current and future scholarly contexts. This opportunity is especially attractive and potentially useful, given the limited cross-fertilisation to date between linguistic and literary/cultural cognitive approaches, and the entrenched predisposition of Anglo-Saxonists for interdisciplinarity and excellent track record in that respect, especially across the literary/cultural and linguistic divide. Furthermore, our field affords deep diachronic access to those scholars considering the interaction of culture and cognition, cross-cultural intelligibility in a very broad sweep, and the interplay of cultural transmission and distributed cognition. Anglo-Saxon Studies is already familiar with the historicised text and vernacular mentalities, and so is poised to embrace cognitive studies of literature and culture in all their manifestations.

This study is a contribution to that project. This book is concerned with the psychological foundations of Anglo-Saxon literary culture, and the cognitive

96 See, for example, Joyce Hill, 'Methodologies, Mantras, and Paradigms: Research in Early Medieval English Literature', *Journal of English and Germanic Philology* 105 (2006), 87–101.

97 See *Parker Library on the Web*, http://parkerweb.stanford.edu; *Early Manuscripts at Oxford University*, http://image.ox.ac.uk; the Bodleian Digital Texts Series; and the recently published *Historical Thesaurus of the Oxford English Dictionary* mentioned above, n. 63.

98 Other possibilities include the Wheaton College Lexomic Group, which analyses discursive strings of letters via computational and statistical analysis. *Wheaton College Lexomics*, http://lexomics.wheatoncollege.edu.

technicalities of textual interpretation by both contemporary and remote audiences. It seeks to deploy ideas and techniques from a range of cognitive approaches to literature and culture in order to account for how twenty-first-century readers are able to create meaning from literary texts that are over one thousand years old.

Chapter 2 starts this process by focusing on a key element of many cognitive approaches: metaphor. This chapter traces and adapts for the Anglo-Saxon context the influential conceptual metaphor theory developed by Lakoff and Johnson, with a particular focus on the idea of how embodied experience determines metaphor clusters in relation to the life journey and the mind. This investigation takes a particular interest in the metaphors of the wandering mind in *The Wanderer* and the body as an enclosure in *Soul and Body II*. It closes with a consideration of embodied cognition.

Chapter 3 develops Mark Turner's idea of conceptual blending, central to cognitive studies of metaphor and literary meaning, and then applies that theory to three distinctly different Old English poetic texts: *The Dream of the Rood*, *Riddle 43* and an excerpt from *The Battle of Maldon*.

Chapter 4 develops Text World Theory, a core idea from Cognitive Poetics that considers the way in which contextual realities are introduced into the process of interpretation by the textual recipient. After outlining this interpretive technique, this chapter demonstrates it in practice in relation to excerpts from *Beowulf* and *Genesis B*, and the whole short poem, *Wulf and Eadwacer*.

Chapter 5 focuses on the idea of mind-reading or Theory of Mind within the approach now known as Cognitive Cultural Studies. This chapter demonstrates the distinctively different approach offered by this school of thought by revisiting a text treated in Chapter 3, *The Dream of the Rood*, and one from Chapter 4, *Beowulf*, as well as a new one here, *Elene*. After these demonstrations of the theoretical ideas, the chapter closes with a consideration of the idea of cognition as a cultural process in Anglo-Saxon England.

Chapter 6 explores the reliance of Cognitive Science on narrative, especially in relation to memory and the development of the self, and explores the potential of a fuller collaboration of this field with Literary Studies in general, and Anglo-Saxon Studies in particular. It considers the capacity of poetry to represent subjective states and autobiographical memory, and thereby to enact the dynamic interaction of culture and cognition. The chapter takes as its focus on a text that apparently relies heavily on autobiographical narrative, *The Wife's Lament*.

Chapter 7 reconsiders two texts treated above – *Wulf and Eadwacer* and *Beowulf* – from a new perspective, a cognitive history of emotions. It demonstrates the value of combining an established approach to literary history with the contribution provided by consideration of emotions made available by Cognitive Science.

The Conclusion draws the discussion together with an overview of the web of associations activated by the language and underlying conceptual structures of vernacular texts, and considers the way in which cognitive approaches can enhance and supplement, rather than replace, well-established means of exploring Old English texts. Finally, the discussion turns to observations on a specifically cognitive way forward for Anglo-Saxon Literary Studies through the range of perspectives made available by emerging cognitive approaches, and in turn what the long range-view of Anglo-Saxonists can offer Cognitive Science.[99]

[99] An earlier version of portions of this chapter appeared in 'Cognitive Studies of Anglo-Saxon Mentalities', *Parergon* 27.1 (2010), 13–26.

2

Conceptual Metaphors

Theories of conceptual metaphor

Metaphor is the common factor in all cognitive approaches to the study of literature, so is the natural starting point for considering culture, cognition and literature. Traditionally, a metaphor has been considered a linguistic expression where the attributes of the source domain are transferred onto those of the target domain. For instance, in the expression 'life is a dream', selected attributes of the concept 'dreaming' are mapped onto those relating to the idea 'life', resulting in an innovative synthesis. Recently, though, findings from cognitive science have demonstrated that such uses of language are the surface manifestations of deep-seated human mental processing that is fundamentally metaphorical. Many branches of Cognitive Science have established that everyday mindwork relies on mapping the characteristics of one conceptual domain onto another, and that the product is more than the sum of the parts, or more than a simple blend or mapping. Our capacity for abstract reasoning, imagination, learning and comprehension all depend on being able to blend ideas, reconceptualise one thing in terms of the characteristics of another, make mental representations of the combination, and draw analogies from the similarities and differences. Taken beyond the linguistic into the cognitive sphere, metaphor has therefore become the subject of crucial interest to literary and linguistic cognitive approaches, as well as being the centre of attention and debate more broadly in a wide variety of disciplines. The sheer magnitude and vigour of current discussions are indicated by the number of items in a current online bibliography,[1] and the huge surveys of key theories in particular disciplines such as Philosophy and Linguistics.[2]

This shift in the way metaphor is seen, and its growing significance for a wide range of cognitive approaches, was galvanised by the view that metaphors are conceptual as well as linguistic: we use metaphors not only

1 There are 6500 items from 1990 onwards to be found in Sabina De Knop, René Dirven, and Birgit Smieja, eds, *Bibliography of Metaphor and Metonymy* (an online resource from John Benjamins Publishing Company), accessed 31 August 2011 at: http://www.benjamins.com/online/met/. For a diachronic overview of the development of metaphor theories, see Miriam Taverniers, *Metaphor and Metaphorology: A Selective Genealogy of Philosophical and Linguistic Conceptions of Metaphor from Aristotle to the 1990s* (Gent, 2002).

2 Philosophy: Mark Johnson, ed., *Philosophical Perspectives on Metaphor* (Minneapolis, 1981), with annotated bibliography on pp. 329–52; Linguistics: Wolf Paprotté and René Dirven, eds, *The Ubiquity of Metaphor: Metaphor in Language and Thought* (Amsterdam and Philadelphia, 1985).

to express ourselves in language, but also to formulate ideas and to think.[3] The idea that metaphor plays a fundamental role in structuring our thought and language, through mapping embodied and spatial experiences onto abstract concepts, was put forward by George Lakoff and Mark Johnson in *Metaphors We Live By* and subsequent publications.[4] Their theory positions metaphors as cognitive models that express deep-seated mental ways of organising thought, enacted in language through individual projections (p. 18). For example: THE MIND IS A CONTAINER is a conceptual metaphor that underpins individual expressions, or linguistic metaphors, such as 'the idea came into my mind' and 'file away that idea'. Other conceptual metaphors for the mind include THOUGHT IS OBJECT MANIPULATION ('to grasp an idea') and MENTAL PROCESSING IS INGESTION ('digest the contents of a book'). Lakoff and Johnson show convincingly that such metaphors are part of stable and productive conceptual systems that underpin everyday communication, and that these metaphors thereby organise personal and social experience: they condition our thought processes and systematically influence behaviour. That is, people actually think in terms of the mind being a container and use that overarching idea to conceptualise, organise and access their memories, and in turn to create and decode individual metaphoric expressions from this basis in everyday communication (p. 61). This is how metaphors that rely on this systematic mapping of ideas can be comprehensible to another person. The superordinate metaphors, such as LIFE IS A JOURNEY and THE MIND IS A CONTAINER, implicitly give rise to limitless individual metaphors such as 'to go off the straight and narrow' and 'to have a closed mind', which would otherwise be meaningless or inscrutable.

Lakoff and Johnson's pragmatic approach to language use is helpful for understanding how metaphor works, how influential it is in shaping and organising our mental lives, and how it is created at a conceptual rather than a linguistic level. Their work has been hugely influential in promoting conceptual metaphor into a place of central significance in several disciplines, and in turn has laid the foundation for a new field, Cognitive Linguistics, and has influenced the development of many cognitive approaches to literature. This work is so widely cited that almost anyone in any field who discusses metaphor mentions 'Lakoff and Johnson', and takes the existence of conceptual metaphor, and the ubiquity of conventional metaphors in everyday language, as givens.[5]

But while there is no doubt that the idea of pervasive and systematic conceptual metaphors is inherently valid, there are some major limitations to their method that urge caution for the Anglo-Saxonist intent on applying

3 For example, Raymond W. Gibbs Jr, *The Poetics of Mind: Figurative Thought, Language and Understanding* (Cambridge, 1994).

4 George Lakoff and Mark Johnson, *Metaphors We Live By* (Chicago, 1980); see also *Philosophy in the Flesh: The Embodied Mind and its Challenge to Western Thought* (New York, 1999). On the impact of conventional metaphor clusters on the expression of emotion, see also Lakoff and Johnson, *More than Cool Reason: A Field Guide to Poetic Metaphor* (Chicago and London, 1989).

5 See the introduction in Antonio Barcelona, ed., *Metaphor and Metonymy at the Crossroads: A Cognitive Perspective* (Berlin and New York, 2000), pp. 1–28.

their ideas. For example, Lakoff and Johnson, who concentrate mainly on present-day spoken English, do not address idiosyncratic or innovative metaphor use, such as is characteristically a feature of 'literary' technique and style; and they do not look at the origin, development or transmission of these metaphors or categories over time. These omissions are problematic, as there is some evidence to suggest that metaphors are inherently 'literary' by nature if not in use (in that literary language defamiliarises experience and draws attention to itself), and that metaphoric categories in use today have probably been in use, at least in English, for some considerable time.[6] Lakoff and Johnson therefore do not consider two vital factors relating to human communication: the structure of knowledge at a cultural level, and the impact of cultural transmission and diachronic development of metaphor use.[7] An increasing number of critics have found other inadequacies in the theory, which has come to be characterised as sweeping and vague, and certainly in need of refinement, notwithstanding its evident impact and intuitive aptness.[8] Most importantly, the discourse context of metaphor must be taken into account in the interpretation of the expression (particularly important for recurring metaphors or those that develop within a larger piece of text, rather than in a single expression). The literary context (especially the poetic one) provides some further complicating features that prevent Lakoff and Johnson's findings from being generalised to this domain. When readers know they are reading literary texts, they mobilise special reading strategies and knowledge about literary discourses, which guide their perception processes and their interpretations of metaphor.[9]

Such considerations, at the heart of Cognitive Literary/Cultural Studies and Cognitive Poetics, are crucial for understanding fully the contribution of conceptual metaphor to meaning, and therefore contextualised metaphor is of key interest to scholars working in these fields. So, notwithstanding its perceived limitations and perhaps generalisations, the rich seam of inquiry opened up by the work of Lakoff and Johnson has been instrumental in encouraging cross-disciplinary studies of metaphor, in particular the application of Cognitive Linguistics to the study of literature.[10] Scholars have

6 Tracing the biographies and extra-literary migrations of such metaphors is my larger ongoing research project.

7 For other objections, based on metaphor complexity and contextual factors, see Michiel Leezenberg, *Contexts of Metaphor* (Amsterdam, 2001), pp. 136–7; and David L. Ritchie, *Content and Connection in Metaphor* (Houndmills, 2006), pp. 40–2.

8 For a sustained attack on Lakoff and Johnson's lack of a philosophy of language, and their 'inconclusive arguments', as well as details of other criticism, see Verena Haser, *Metaphor, Metonymy and Experientialist Philosophy: Challenging Cognitive Semantics* (Berlin and New York, 2005). Notwithstanding her wide-ranging criticisms, most notably the contention that linguistic metaphors can spring from more than one conceptual metaphor, Haser concedes 'the undeniable merits of the conceptual metaphor view' (p. 240). She proposes an alternative theory, based on family resemblances or analogies among figurative expressions, and emphasises the need for the linguistic context of individual expressions.

9 See Gerard Steen, *Understanding Metaphor in Literature: An Empirical Approach* (London and New York, 1994), esp. p. 47.

10 Mark Turner, *Reading Minds: The Study of English in the Age of Cognitive Science* (Princeton, 1991), makes the case for the way Literary Studies can be revolutionalised by a consideration of

recently brought cultural and communication theory to the study of mind and metaphor, thereby expanding the range of considerations within our understandings of metaphor use and interpretation.[11] The arguments of Lakoff and Johnson have been extremely influential in the investigations being made in cognitive studies of literature. In particular, they argue that our understanding and communication of human existence arises from fundamental acknowledgement of existence within a physical body, which in turn provides a basis for all kinds of conceptual metaphors.[12] The concept of embodiment has been taken up by scholars enthusiastically, including in later work by Lakoff and Johnson themselves, who argue that material embodiment provides an inescapable context of meaning, and in turn motivates the creation of metaphors experientially. Lakoff and Johnson use the example of upright human posture and the embodied feeling of gravity to explain the orientational conceptual metaphor UP IS GOOD ('keep your spirits up', 'an upstanding person').[13] The key concept of embodiment, then, has become prominent in cognitive literary discussions, with particular relevance to metaphor.

These ideas can be applied fruitfully to develop our understanding of mental schemas that inform the representations of abstract ideas such as consciousness in earlier literary texts, and to make cross-cultural and diachronic comparisons of mental models in order to shed light on the impact of culturally specific factors on embodied and semantic universals.[14] Recent research into mental processes in Cultural Psychology suggests that the way the mind and its attributes and functions are understood is culturally specific to a high degree and varies across time and among societies, because cognition is in part determined by social experience.[15] Studies of the histories

reading and writing as cognitive activities. Similarly, Zoltán Kövecses, *Metaphor: A Practical Introduction* (Oxford, 2002), brings together the study of language, culture and the mind in a cognitive-linguistic theory of metaphor, and collections of studies provide multiple perspectives on the cognitive approach to metaphor. See, for example, Raymond W. Gibbs Jr and Gerard Steen, eds, *Metaphor in Cognitive Linguistics* (Amsterdam and Philadelphia, 1999), and Barcelona, *Metaphor and Metonymy*.

11 Naomi Quinn, for example, argues that 'cultural understanding underlies metaphor use' through the 'organizing role of culture in human thought'. See 'The Cultural Basis of Metaphor', in *Beyond Metaphor: The Theory of Tropes in Anthropology*, ed. James W. Fernandez (Stanford, 1991), pp. 56–93, at p. 57. Similarly, the link between the use of figurative language and ideology is the key focus of Jonathan Charteris-Black in *Corpus Approaches to Critical Metaphor Analysis* (Houndmills, 2004).

12 Lakoff and Johnson, *Philosophy in the Flesh*.

13 Lakoff and Johnson, *Metaphors We Live By*, pp. 14–21.

14 See, for example, Caroline Gevaert, 'The Anger is Heat Question: Detecting Cultural Influence on the Conceptualisation of Anger through Diachronic Corpus Analysis', in *Perspectives on Variation: Sociolinguistic, Historical, Comparative*, ed. N. Delbecque, J. van der Auwera and D. Geeraerts (Berlin and New York, 2005), pp. 195–208; Marie-Odile Junker, 'A Native American View of the "Mind" as Seen in the Lexicon of Cognition in East Cree', *Cognitive Linguistics* 14.2/3 (2003), 167–94; and Richard E. Nisbett, *The Geography of Thought: How Asians and Westerners Think Differently . . . and Why* (New York, 2003), esp. pp. 39–45.

15 See Petra Hauf and Friedrich Försterling, *Making Minds: The Shaping of Human Minds through Social Context* (Amsterdam and Philadelphia, 2007), esp. p. 2: 'individual minds become social minds by learning to share their mental contents with others'. See also R. A. Shweder and Robert A. LeVine, eds, *Culture Theory: Essays on the Mind, Self, and Emotion* (Cambridge, 1984);

of ideas about the mind are also shedding light on the way culture and mind schemas interact.[16] Similarly, the linguist Zoltán Kövecses emphasises the impact of culture on metaphor production and variation.[17] His conclusion, that 'metaphor is inherently *conceptual, linguistic, neural-bodily*, and *social-cultural* – all at the same time',[18] captures the complexity of the topic and its embeddedness in lived experience. His claim that both the universality and variation of metaphors rely on embodiment, socio-cultural context and individual cognitive preferences is comprehensively convincing, and ought to be accommodated in any serious study of mind and metaphor.

Metaphor, then, has come to be seen as a cognitive process, as well as a ubiquitous and a unique form of communication, which must be examined in a fundamentally transdisciplinary way that embraces mental as well as literary processes. As Yeshayahu Shen, the editor of an important collection of articles devoted to the subject, argues: metaphor is not simply a phenomenon of language but rather a 'much more widespread conceptual process which should be examined from a broader interdisciplinary perspective'.[19] As a conceptual phenomenon, metaphor insists on some degree of cognitive analysis such as is afforded by the discipline Cognitive Psychology, with its focus on thinking, reasoning and remembering.[20] In the collection edited by Shen, the psychologist Raymond W. Gibbs claims, in his broad-ranging analysis of the reasons that theories of metaphor vary so widely, that theories focus on different moments in the process of interpretation.[21] The function of metaphor is similarly privileged by the psychologists Boak Keysar and Sam Glucksberg, who distinguish metaphor from other types of communication on the basis of how it is used rather that the way it is comprehended.[22]

Notwithstanding this sustained and intense scrutiny of metaphor, literary and other contexts are rarely afforded as much attention as the structure and comprehension of metaphor itself. Somewhat surprisingly, at least from the perspective of the literary scholar, the discourse context of poetry is rarely a matter of interest to those who study metaphor (even to the majority who refer to this trope as a feature of 'poetic' language), and only relatively recently has the impact of culture on cognition and on the diachronic ways of imagining and expressing ideas about that cognition started to receive any consideration.

R. D'Andrade, 'A Folk Model of the Mind', in *Cultural Models in Language and Thought*, ed. D. Holland and N. Quinn (Cambridge, 1987), pp. 112–48; and R. A. Sharpe, *Making the Human Mind* (London, 1990).

16 Paul S. Macdonald, *History of the Concept of Mind: Speculations about Soul, Mind and Spirit from Homer to Hume* (Aldershot, 2003).

17 Zoltán Kövecses, *Metaphor in Culture: Universality and Variation* (Cambridge, 2005).

18 Ibid., p. 293.

19 Yeshayahu Shen, 'Cognitive Aspects of Metaphor Comprehension: An Introduction', *Poetics Today* 13.4 (1992), 567–74, at p. 568.

20 For a useful introduction that explores the connection with Philosophy, see Rom Harré, *Cognitive Science: A Philosophical Introduction* (London, 2002).

21 Raymond W. Gibbs Jr, 'When is Metaphor? The Idea of Understanding in Theories of Metaphor', *Poetics Today* 13.4 (1992), 575–606, at p. 602.

22 Boak Keysar and Sam Glucksberg, 'Metaphor and Communication', *Poetics Today* 13.4 (1992), 633–58, at p. 656.

As Dirk Geeraerts and Stefan Grondelars insist, cultural history has an impact on both language and cognition, thereby allowing metaphor usage a key role in the history of ideas.[23] Draaisma's survey of metaphors for the mind from Plato to the present day makes the same case, arguing for the 'heuristic function of metaphor in terms of influencing interpretation as a filter'.[24] The study of metaphor creation and transmission, then, can be one component of a cognitive historicist approach to literary meaning, in which scholars 'address the complex interrelation of evolved neurocognitive structures and contingent cultural environments with an eye to specific examples of cultural change'.[25]

Scholars working within a wide range of cognitive approaches, however, are making advances of direct relevance for the literary scholar, by focusing on the contextualised deployment of conceptual metaphors in literary texts, and the way they operate within systematic thought processes rather than as an isolated language feature.[26] Margaret H. Freeman outlines the case for this sort of project from the perspective of Cognitive Linguistics, where systematic cross-domain mappings account for the occurrence and patterning of metaphors underpinning an interpretive response to a literary text.[27] Freeman approaches literature as the product of a cognising mind, with analogical reasoning, the capacity to integrate concepts, and create and interpret systematic mappings of attributes that generate information and connections beyond what is on the page. Similarly in Linguistics, F. Elizabeth Hart attempts to identify this difference between literary and other metaphor as one of degree or richness: 'literary metaphor extends and elaborates by expanding the bounds of a culturally conventional metaphor to include more aspects than is conventionally necessary'.[28] Likewise, Gerard Steen distinguishes complex from simple metaphor in his study of the impact of special discourse context features of literary texts on linguistic processing.[29] In Cognitive Stylistics, Elena Semino focuses separately on creativity, variation and impact in literary, political, scientific and other uses of metaphor.[30] Her conclusion, that literary metaphors rely on creative variations on conventional metaphors,[31] provides further support for Lakoff and Johnson's initial claim, and specific textual instances of their propositions, though lays greater stress

23 Dirk Geeraerts and Stefan Grondelars, 'Looking Back in Anger: Cultural Traditions and Metaphoric Patterns', in *Language and the Cognitive Construal of the World*, ed. John R. Taylor and Robert E. MacLaury (Berlin and New York, 1995), pp.153–79.

24 D. Draaisma, *Metaphors of Memory: A History of Ideas about the Mind*, trans. Paul Vincent (Cambridge, 2000), p. 18.

25 Alan Richardson and Francis F. Steen, 'Literature and the Cognitive Revolution: An Introduction', *Poetics Today* 23.1 (2002), 1–8, at p. 5.

26 See the overview in M. Fludernik, D. C. Freeman and M. H. Freeman, 'Metaphor and Beyond: An Introduction', *Poetics Today* 20.3 (1999), 383–96.

27 Margaret H. Freeman, 'Poetry and the Scope of Metaphor: Toward a Cognitive Theory of Literature', in *Metaphor and Metonymy*, ed. Barcelona, pp. 253–82.

28 F. Elizabeth Hart, 'Cognitive Linguistics: The Experiential Dynamics of Metaphor', *Mosaic* 28.1 (1995), 1–23, at p. 15.

29 Gerard Steen, 'Analyzing Metaphor in Literature: With Examples from William Wordsworth's "I Wandered Lonely as a Cloud"', *Poetics Today* 20.3 (1999), 499–522.

30 Elena Semino, *Metaphor in Discourse* (Cambridge, 2008).

31 Ibid., p. 80.

on the rhetorical and aesthetic potential of patterns of creative metaphor usage in authentic contexts, a line of inquiry that can certainly be developed further in relation to other literary texts. From the Cognitive Literary Studies perspective, the role of conceptual metaphor in complex literary meaning-making, and the potential of the study of metaphoric systems to provide traces of earlier ways of thinking, is an avenue of research taken up particularly by early-modern scholars, including Mary Thomas Crane.[32] From the Cognitive Poetics perspective, Paul Werth's idea that literary metaphor is distinctive in its use of motivated linguistic choices and sustained or recurrent metaphors, or megametaphors,[33] has been taken up and developed by other scholars, most notably by Peter Stockwell.[34] All cognitive approaches consider the complex mappings of conceptual metaphors in discourse, and scholars like Stockwell are leading the way in considering the particularly dense clustering and patterning of metaphors in literary texts.

These and other discussions have demonstrated the fruitful combination of the literary with the cognitive in the study of metaphor, and have underpinned all cognitive approaches to Literary Studies, which continue to centre on conceptual metaphor. In particular, the growing acknowledgement of the impact of embodied experience on language use and the cognitive reliance on conceptual blending have demonstrated how central metaphor is to human experience and communication. As such, understanding the systematic generation and clustering of metaphors can be a valuable tool in cultural and textual understanding, especially considering the continuity of embodied experience alongside cultural variability in human history. The potential of this fundamental cultural, linguistic and cognitive phenomenon for further historical and diachronic scrutiny has become increasingly appealing for scholars of the Middle Ages, though as yet only in certain confined areas with respect to Anglo-Saxon England.

Old English conceptual metaphors for life, body and mind

Conceptual metaphor use in Old English poetry, particularly the habit of conceptualising the mind as a figurative container for thoughts, has received some attention recently, though there is clearly a great deal more to be done in this area.[35] In particular, the presence of this metaphor alongside the figurative

32 Mary Thomas Crane, 'Analogy, Metaphor, and the New Science: Cognitive Science and Early Modern Epistemology', in *Introduction to Cognitive Cultural Studies*, ed. Lisa Zunshine (Baltimore, 2010), pp. 103–14.

33 Paul Werth, *Text Worlds: Representing Textual Space in Discourse* (London, 1999), p. 317–23.

34 Peter Stockwell, *Cognitive Poetics: An Introduction* (London, 2002), p. 111.

35 Most recently Leslie Lockett, *Anglo-Saxon Psychologies in the Vernacular and Latin Traditions* (Toronto, 2011), developed out of her thesis, 'Corporeality in the Psychology of the Anglo-Saxons', Ph.D. thesis, University of Notre Dame, 2004, DAI-A 65/06 (2004); Britt Mize, 'The Representation of the Mind as an Enclosure in Old English Poetry', *Anglo-Saxon England* 35 (2006), 57–90; and Mize, 'Manipulations of the Mind-as-Container Motif in *Beowulf, Homiletic Fragment II*, and Alfred's *Metrical Epilogue to the Pastoral Care*', *Journal of English and Germanic Philology* 107 (2008), 25–56. Both Lockett and Mize build on the ideas articulated by Soon-Ai

and literal conceptions of the body as a container complicates the picture, as does the presence of other very conventional metaphors, many of which were imported into Anglo-Saxon England along with Latin literacy and Christianity. Some aspects of metaphor deployment were inherited with Latin texts, where figurative expression offers a means of capturing otherwise inexpressible conundra. Medieval Christian writers were especially fond of the metaphor as a tool for reflecting on and explaining the mysteries of faith. The definition, categorisation and discussion of metaphor discussed by Aristotle and within the rhetorical treatises of the Stoic tradition were adopted and adapted during the Middle Ages, most notably in the key statement on the value of this trope and others in Christian literature in Augustine's *De Doctrina Christiana*.[36]

This fondness for metaphor was adopted and adapted in Anglo-Saxon England, as well as a native fluency in metaphor creation and conceptual blending. In extant Old English poetry, three particularly prevalent metaphors occur, the first two of which are also very common in Latin literature: LIFE IS A JOURNEY, THE MIND IS A CONTAINER and THE MIND IS A WANDERING ENTITY. The third is an interesting addition in this context because a version of it appears in variations in Latin texts, and it is combined in Old English literature with the other two conceptual metaphors in interesting ways. Just how and why the combination is unusual and illuminating is explored below, pp. 48–50. Because these culturally privileged metaphors are widely used in poetic texts, and are blended together into a complex cluster of ideas in the most memorable accounts of the life of the mind, they constitute the focal conceptual metaphors in the following discussion, first of *Soul and Body II*, and then of two of the best-known and most widely studied poetic texts, *The Wanderer* and *The Seafarer. Soul and Body II* provides a complication of THE MIND IS A CONTAINER metaphor in its use of literal and figurative references to the body as a container; the latter two texts have been chosen as they provide instances of complex poetic representations of cognitive functioning. In all three poems, these overarching metaphors are deployed within sustained reflections on the transitory corporeal life. In *The Wanderer* and *The Seafarer*, the mind is imagined as both a storage place for thoughts and also something that can wander away. At first blush, this combination might seem incongruous, yet these figurative schemas underpin the conceptualisation of mental life in a wide range of Old English poetry. In emphatic and poetically complex combination in these texts, they form a dual model of the containing, wandering mind within a vernacular poetic psychology that privileges the collection and storage of wise thoughts, judicious verbal expression and motivated self-

Low, 'The Anglo-Saxon Mind: Metaphor and Common-Sense Psychology in Old English Literature', Ph.D. thesis, University of Toronto, 1998, DAI-A 60/10 (2000); Soon-Ai Low, 'Approaches to the Old English Vocabulary for "Mind"', *Studia Neophilologica* 73 (2001), 11–22; Soon-Ai Low, 'Mental Cultivation in *Guthlac B*', *Neophilologus* 81 (1997), 625–36; and Michael Matto, 'Containing Minds: Mind, Metaphor and Cognition in Old English Literature', Ph.D. thesis, New York University, 1998, DAI 59A (1999).

36 M. Irvine, *The Making of Textual Culture: 'Grammatica' and Literary Theory 350–1100* (Cambridge, 1994), pp. 104 and 107.

control. The following study of these three poems demonstrates a cognitive approach of particularly broad application afforded via considerations of conceptual metaphor in three specific instances.

The body as an enclosure in Soul and Body II

Somewhat counter-intuitively, often the most conventional conceptual metaphors and schemas can be the most interesting and the most illuminating, though the least studied, especially in relation to literary language and meaning. While THE MIND IS A CONTAINER metaphor has received some attention lately, the idea of THE BODY IS A CONTAINER has received far less, possibly because the body can literally be considered a container (of organs, biological systems, blood and food) as well as figuratively, though the two operate together in Anglo-Saxon England as they do in other cultures. Indeed, the literal truth of the body as a container makes the idea of metaphor complex or even questionable in some instances. The Exeter Book poem known as *Soul and Body II* provides a good example of how these conventional metaphors act in concert, and how conceptual theories can elucidate the underlying patterns of thought informing a literary text. In this poem, THE BODY IS A CONTAINER is literalised for rhetorical effect, and the double vision made possible by the conceptual metaphors of the body as an enclosure, prison garment and house creates a complex set of associations that play out rhetorically and aesthetically in the poem.

In *Soul and Body II*, the soul returns to the body after death to berate it for its former fleshly pleasures and to warn of the horrible process of consumption by worms and decomposition that awaits the material corpse. This theme is common in medieval literature and beyond,[37] and indeed informs the very similar Old English poem, *Soul and Body I* in the Vercelli Book, with which our focal text might share an exemplar, though from which it deviates in its omission at the end of the account of the blessed soul.[38] The polemical function of *Soul and Body II* is quite evident and undisputed – the body undisciplined in life will imperil the fate of the eternal soul – though readers see different

37 Robert W. Ackerman, 'The Debate of the Body and the Soul and Parochial Christianity', *Speculum* 37.4 (1962), 541–65; Piero Boitani and Anna Torti, eds, *The Body and the Soul in Medieval Literature* (Cambridge, 1999); Rosalie Osmond, *Mutual Accusation: Seventeenth-Century Body and Soul Dialogues in their Literary and Theological Context* (Toronto, 1990); and Helena Znojemská, 'Medieval English Soul and Body Literature', *Litteraria Pragensia: Studies in Literature and Culture* 17.34 (2007), 20–39. See also John Justin Brent, 'The Legend of Soul and Body in Medieval England', Ph.D. thesis, State University of New York, Stony Brook, 2000, DAI-A 61/09 (2001); and Eve E. Siebert, 'Body and Soul Poems in Old and Middle English', Ph.D. thesis, Saint Louis University, 2008, DAI-A 70/03 (2009).

38 P. R. Orton, 'The OE "Soul and Body": A Further Examination', *Medium Ævum* 48 (1979), 173–97; and Douglas Moffat, 'The MS Transmission of the OE *Soul and Body*', *Medium Ævum* 52 (1983), 300–2. Michael Matto, 'The Old English *Soul and Body I* and *Soul and Body II*: Ending the Rivalry', *In Geardagum* 18 (1997), 39–58, argues that it is most useful to consider the poems separately. For detailed discussions of the sources, analogues, criticism up to 1990, and an edition and translation of both poems, see Douglas Moffat, *The Old English Soul and Body* (Wolfeboro, NH, 1990).

injunctions and anxieties embedded within it. Many impute an orthodox eschatological message to the text, including a symbolic function to the named parts of the body, indicating separate punishment in hell.[39] Allen J. Frantzen has refined that idea further by arguing that the poem relies on the penitential metaphor of sins as wounds to explain its theological framing.[40] More recently, Glenn Davis has claimed that, given its emphasis on the fragmentation of the body, the poem reflects concern for physical health and welfare rather than the afterlife,[41] though this argument is less convincing than others. Michelle Hoek's idea, that the vivid depictions of physical decay in this poem invert the Platonic metaphor of the body as prison of the soul, is even more interesting.[42] Hoek explores the presentation of decay in the poem as a representation of the breaking down of the distinction between the inside and outside of the body, particularly given the emphasis made in the poem on the mouth and its main functions of speaking and eating, both of which involve transmission across both sides of the internal and external regions.[43] But even more pertinent than these arguments is the distinction drawn by Malcolm Godden between the Platonic concept of the tripartite soul that informs Old English prose, and the fundamental distinction between the mind (controlling thought, emotion and volition) and the soul that informs the verse.[44] This distinction seems to be upheld in *Soul and Body II*, as Godden points out,[45] in that the soul, freed from the body after death, is made to suffer for the wrongs committed by the psychological faculty, the mind. This schema of a split self would certainly account for the way in which the reproach is framed within the poem, and explain the reason for the soul's expressed grievance with the body-based mind.

These critical readings, like the poem itself, rely on implied conceptual metaphor, THE BODY IS A CONTAINER. This metaphor is so commonplace as to pass unnoticed, yet what is fascinating in respect of this poem is the overlap between the figurative and literal ideas of the body as a container: of the soul, of desire, of organs and of flesh-eating worms. In this poem, the behaviour in life and then decomposition in death of the physical body are of central interest. Metaphors of enclosure are particularly prominent. The voices of both the narrator and the soul refer to the body as a variety of different yet conventional containers. It is a dwelling that the soul shares with raging desires: 'eardode ic þe in innan' (30a), 'I lived within you', it says, 'ond me firenlustas / þine geþrungon' (31b–32a), 'and your wicked desires thronged

39 Thomas D. Hill, 'Punishment According to the Joints of the Body in the Old English "Soul and Body II"', *Notes and Queries* 15 (1968), 409–10, and 'Punishment According to the Joints of the Body, Again', *Notes and Queries* 16 (1969), 246.

40 Allen J. Frantzen, 'The Body in *Soul and Body I*', *Chaucer Review* 17 (1982), 76–88.

41 Glenn Davis, 'Corporeal Anxiety in Soul and Body II', *Philological Quarterly* 87.1–2 (2008), 33–50.

42 Michelle Hoek, 'Violence and Ideological Inversion in the Old English *Soul's Address to the Body*', *Exemplaria* 10 (1998), 271–85.

43 Ibid., pp. 279–81.

44 Malcolm Godden, 'Anglo-Saxons on the Mind', in *Learning and Literature in Anglo-Saxon England*, ed. M. Lapidge and H. Gneuss (Cambridge, 1985), pp. 271–98, at p. 289.

45 Ibid., p. 289.

about me'. The earlier containment of the soul by the body is repeated at line 40 'ic þe in worulde wunian sceolde', 'I had to exist within you in the world', and again at line 66, where the body is called a 'dishonourable dwelling place', 'þa arleasan eardungstowe'. Besides many other references to having been 'within' the body (e.g. 43a, 52a), the body is also described as the clothing worn by the soul: 'þe heo ær longe wæg' (11b), 'that it previously wore', a mapping that is invoked at the conclusion of the poem when the body is referred to as 'þæt he longe ær / werede mid wædum'(118b–119a), 'that which previously was long clothed with garments'. These metaphors are cross-mapped with others in which the body is the kinsman of the soul sundered by death (line 4), then dust (16, 17, 99), clay (19), filth (18), decay (18) and finally as food for worms (21, 67–9, 106–8, 119). This last metaphor entails a further cross-mapping, of the container becoming the contents of a new container, the worm, once the body ceases to contain the soul. This metaphor is central to the eschatological thrust of the poem, as it is elaborated at the conclusion in the description of the lead worm's penetration of the cranial cavity (*Soul and Body II*, 112b–117a):

Se geneþeð to
ærest ealra   on þam eorðscræfe;
he þa tungan totyhð   ond þa toþas þurhsmyhð,
ond to ætwelan   oþrum gerymeð,
ond þa eagan þurhiteð   ufon on þæt heafod
wyrmum to wiste.

He ventures forth first of all into the sepulchre, slits the tongue and bores through the teeth, and enlarges it for the others for the feast, and penetrates the eyes up into the head for the worms as a banquet.

In this development of the BODY IS A CONTAINER metaphor, the head (itself the container of the mind, or at least the physical brain in this worldview) becomes a literal container for worms: a sepulchre (113). The poem expresses the message that once the body becomes a literal container, all is lost, including the status of container as the body becomes contained in the most ignominious, revolting way. The most personal, durable, and potentially elevated contents of the mind – thoughts, beliefs, memories, and even some qualities of the self or the soul – have been replaced by the lowest, most mundane, and alien physical contents: feasting worms. The abstract idea has been reduced to the most physical of substances, food.

The soul is presented as the now-removed contents of this container, responsible for the integrity of the whole: it says the body is 'gestaþelad þurh mec' (42b), 'stabilised by me'. The soul is now a different entity altogether from the way it was when inside the living body: it has freedom of movement, powers of expression, and a capacity for scorn aimed at its incomplete autonomy. In this scenario, the body retains its 'container' function, which has become literalised – it now houses worms in place of the soul. In a series of reversals and paradoxical statements, the soul-container, once empty, becomes meaningless, the ravenous body has become food for worms, the feast empty, and the garment of the soul itself is without clothing. The silenced and now consumed tongue, and the way in which the fleshly appetites have caused the flesh itself to be eaten, vividly support the soul's contention that the body is reaping the consequences of being an inadequate home, or container, during life.

The double vision enabled by metaphor allows this conceptual richness and polemical punch, especially in the conclusion of the poem, where the reader is admonished to bear its images in mind: to fill the figurative container-mind with images of the container-body. The concluding lines exhort: 'þæt mæg æghwylcum / men to gemyndum modsnotterra' (120b–121), 'every man wise in mind may recall this'. In the final alliterating stressed syllable of the poem, bringing 'gemyndum' and 'modsnotterra' together, the reader is advised to use the mind as a source of control to regulate the desires of the body, exercise the will, and use the organs of articulation wisely while they are still under one's conscious control. The systematic mapping of body as container, then the substitution of a mind-as-container metaphor, allows a multiple mapping process to occur in the mind of the reader, synthesising literal and figurative contents and containers to make the abstract idea of punishment seem concrete and realistically threatening. The continued relevance and recognition of the conceptual metaphors invoked make the poem not only comprehensible but still powerful today, as modern readers are as likely as any medieval audience to react viscerally to the idea of worms in the cranial cavity. Indeed, as Damasio argues, human emotional reactions, including those evoked by fiction or imagination, have a bodily basis.[46] This means that while precise reactions to imagined horrors may vary in culturally and individually specific ways, the basic human instinct to recoil from projecting such states onto their own bodies transcends these variations.

This poem presents a vividly realised embodied punishment for bad behaviour, but in turn constitutes a form of psychological discomfort in what it invites its audience to do. By presenting an invitation to imagine an observed corpse, it creates a mental representation of and perhaps a visceral reaction to the former living body by contrast – full of sensation, perception, needs, desires and pleasures. The pleasure of being alive, as unlike the corpse as possible, is one implied result of calling up the images represented. Embodiment and the embodied mind are thereby vividly acknowledged and celebrated, while the apparent separation of the soul and its material container, the body, provides an opportunity for asserting their fundamental interaction and interdependence. This reading effect, in the mind of the recipient, is possible through the continued use and transmission of particular metaphors over time, and the similarity of lived, embodied human experience over the millennium between the time when the poems were written down and the present day.

[46] Antonio R. Damasio, *The Feeling of What Happens: Body, Emotion and the Making of Consciousness* (London, 1999), p. 51. See also Chapter 7 below.

The mind as an enclosure

Soul and Body II alludes to one of the most prevalent metaphors in the Western tradition from the early Middle Ages, and a particular favourite of Augustine's[47] – that of the mind as a house or storage space for thoughts and memories, and the imagination.[48] In this spatial conceptualisation, the metaphor THE MIND IS A CONTAINER, mental activity is constructed as travel to, from and within this repository. Anglo-Saxon writers routinely use this metaphor of the mind as a storehouse for ideas and emotions in secular, religious, adapted and apparently original contexts, and poets in particular deploy it flexibly to encompass the concept of containment and escape characterised by various concrete qualities from the natural world. Examples include: a boiling pot or roiling fire (*Seafarer* 10b–11a; *Christ II* 500, 537; *Andreas* 1706); a stream flowing out of a larger body of water (*Wanderer* 53–4); a bird restrained within a cage (*Seafarer* 58–9); or an unspecified enclosure for thoughts: 'hu magon we swa dygle, drihten, ahicgan / on sefan þinne' (*Daniel* 130–131a), 'how, Lord, can we understand (things) so hidden in your mind'. In this latter schema, the mind is believed to have some sort of depth or recessed place that is perhaps secret or hidden, a longstanding commonplace, encoded in poetic compounds that metaphorically capture the idea of thoughts being enclosed in the body: *ferhþcleofa*, 'chamber of the soul, heart, or mind', *ferhþcofa*, 'recess of the soul, heart, or mind', and *ferhþloca*, 'enclosure of the spirit breast'.[49] Embodied thought, contained in the chest-enclosure, is similarly captured in those compounds containing *breost*, which can also connote the mind as a simplex: 'breast as a source of feeling, consciousness'.[50]

Central tenets of this model are the need to store the precious contents of a wise mind securely and to share them in appropriate company sparingly. Often that controlled and contextually charged sharing of words is likened to a ritualistic exchange of gifts bestowing honour on both parties, as is explained in the gnomic poem *Maxims I*.[51] The underlying metaphor, of the

47 St Augustine of Hippo employs many conventional metaphors of the mind as a container or storage receptacle for memories produced by the senses. See B. Stock, *Augustine the Reader: Meditation, Self-Knowledge, and the Ethics of Interpretation* (Cambridge, MA, 1996), p. 217.

48 Of course, there are many other metaphors, both conventional and idiosyncratic, invoked by medieval writers, which we still recognise today. For example, John Cassian favours the mind 'as a great mill-wheel grinding flour for bread'; Bernard of Clairvaux cites 'the stomach of memory'; and many writers, including Bede, used the visual trope of 'the mind's eye', with variations, including the 'spirit's eye' (cf. 'heortan eagum', *Christ I* 1328); see M. Carruthers, *The Craft of Thought: Meditation, Rhetoric, and the Making of Images, 400–1200* (Cambridge, 1998), pp. 91, 95 and 203.

49 *DOE*, s.vv. Cf. *hordcleofa*, 'a treasure-chamber'; *hordcofa*, 'a place for treasure. . .a place where thoughts are stored'; *hordloca*, 'a treasure chest . . . the mind'; *hreþerloca*, 'the breast', *gewitloca*, 'a container for intelligence, the mind', and *wordloca*, 'the storehouse of words' (BT, s.vv.).

50 *DOE*, s. 4. *Breostcofa*, 'breast as a source of feeling, consciousness; figuratively: heart, mind'; *breosthord*, 'heart, mind, literally "treasure of the breast"'; *breostloca*, 'enclosure of the breast, a. figurative: heart, mind'; *breostsefa*, 'heart, mind, innermost thought'; and *breostgehygd*, 'thought of the breast' (*DOE*, s.vv.).

51 Frige mec frodum wordum! Ne læt þinne ferð onhælne, / degol þæt þu deopost cunne! Nelle ic þe min dyrne gesecgan, / gif þu me þinne hygecræft hylest ond þinne heortan geþohtas. / Gleawe men sceolon gieddum wrixlan (*Maxims I* 1–4a) ('Question me with wise words. Do not

mind as a storehouse of words or wisdom that can be unlocked, is at times expressed in emphatic collocations that alliteratively connect wisdom, mind and words: Andreas, 'wis on gewitte, wordhord onleac' (*Andreas* 315), 'wise in mind, unlocked the wordhoard'.[52] More routinely, though, the mind is mentioned as the place in which thoughts, emotions and attitudes are registered: 'flah is geblowen / miclum in gemynde (*Riming Poem* 47b–48a), 'treachery has flourished greatly in the mind'. This wideranging habit of mentioning the place of emotions, and the use of 'in' enshrines a way of conceptualising the mind as a container in everyday as well as poetic language, though in literary contexts, the use of compound words and complex figurative construction amplifies this tendency.

The culturally specific idea that wayward thoughts need to be restrained, and the contents of the mind guarded, is embedded in the conventional collocation of *ferhþ-* compounds with the alliterating adverb, *fæste*, 'firmly'.[53] In *Genesis A*, this collocation is used less characteristically to express the impact of drunkenness on conscious control, in turn conflating mental contents with volition and awareness: 'on ferhðcofan fæste genearwod / mode and gemynde' (2604–5a), '[he was] firmly restricted in the mind's cave, in the mind and the memory'. The latent aspect of mental restraint and escape in these examples shows that Anglo-Saxon writers who use mind-as-container metaphors are reflecting and adapting very widespread, and largely inherited, models. When these schemas are coupled with more local ideas of a peripatetic mind, as they often are, the resulting conceptual blend reflects a uniquely vernacular model of a dualistic embodied mind, and an expression of a genuine attempt to understand and to represent the complexities of human consciousness.

One of the most interesting aspects of the storage metaphor is the corollary notion that the mind has both an inside and an outside, and therefore that thoughts can move in and out of that place. This idea is used widely beyond Anglo-Saxon England, but the transferred notion that the mind itself can travel is not (though the idea of a travelling soul is common in Christian discourse). Partially, this is a matter of terminology, as Old English terms for heart, mind and soul are apparently interchangeable, at least to some degree.[54] But there is evidence to suggest that in Old English poetic texts,

let your mind be concealed, that which you know most deeply to be secret. I will not speak my mystery if you conceal your reason and the thoughts of your heart from me. Wise men must exchange maxims.) On communication as gift exchange, see Robert E. Bjork, 'Speech as Gift in *Beowulf*', *Speculum* 69 (1994), 993–1022. On speech from the chest, see Eric Jager, 'Speech and the Chest: Orality or Pectorality', *Speculum* 65 (1990), 845–59.

52 The clause 'wordhord onleac' occurs on four other occasions, all in poetic texts: again in *Andreas* (601), *Widsith* 1, *Beowulf* 259, and *Metres of Boethius* A6 6.1. Besides containing two of these five instances, *Andreas* also contains the only recorded instance of the compound, *modhord*, 'mindhoard/ treasure of the mind' (172b). Data collected via the *Dictionary of Old English Web Corpus*, ed. Antonette DiPaolo Healey, with John Price Wilkin and Xin Xiang: http://www.doe.utoronto.ca/pub/webcorpus.html

53 *ferhþcleofa* (1 occ. only, with *fæste*); *ferhþcofa* (1 of the 2 occ. with *fæste*); *ferhþloca* (5 of the 11 occasions with *fæste*).

54 Indicated in *DOE* entries cited above, p. 36. The fullest discussion of the vocabulary is

the mind, not the soul, is configured as a travelling entity, that the mind is somehow inside the body but separate from it, and may require some form of control or vigilance by some other entity, perhaps the self.[55] This conflation of mind-as-container schema with that of a wandering or travelling mind that escapes its place of confinement relies on inherited mind schemas that are transmuted in the vernacular context into a paradoxical model of the mind that has a decidedly maritime flavour.

Life as a sea voyage

These mind schemas interact with the related widespread metaphors of life as a sea-voyage, and problems as storms, derive from classical and patristic models, and give rise to many Anglo-Latin and vernacular nautical metaphors, including those relating to anchors.[56] The most prevalent metaphor is the 'ship of state',[57] though the most interesting of these maritime metaphors for our purposes is that of the 'ship of the mind', an idea certainly imported into Anglo-Saxon England with Latin Christianity, though probably pre-existing in vernacular conceptual modes, as both Old English and skaldic poetry show.[58] That vernacular writers seemed to be particularly enamoured of this metaphor is demonstrated in King Alfred's version of Augustine's *Soliloquies*, where, with his usual fondness for metaphoric language, and his propensity for adapting and transforming the figurative language of his sources, the translator adds to the numerous nautical metaphors of the *Soliloquies*.[59] At one point within Book 1, Alfred creates an extended metaphor of the 'ðæt scyp . . . þines modes', 'the ship . . . of your mind'.[60] In view of

still M. J. Phillips, 'Heart, Mind, and Soul in Old English: A Semantic Study', Ph.D. thesis, University of Illinois, 1985, DAI-A 46/07(1986). See also Antonina Harbus, *The Life of the Mind in Old English Poetry* (Amsterdam and New York, 2002), pp. 23–59.

55 Godden, 'Anglo-Saxons on the Mind', p. 294. See also Harbus, *The Life of the Mind*, pp. 137–41.

56 See Roberta Frank, 'Old English *ancor* "anchor": Transformation of a Latin Loanword', in *Germanic Texts and Latin Models: Medieval Reconstructions*, Germania Latina IV, ed. K. E. Olsen, A. Harbus and T. Hofstra (Louvain, 2001), pp. 7–27.

57 Michael McCormack, *Origins of the European Economy: Communications and Commerce, A.D. 300–900* (Cambridge, 2001), p. 404.

58 Frank, 'Old English *ancor* "anchor"': 'in early skaldic verse . . . the breast or heart is a ship, the container of the mind'; and 'the idea of thought or passion as wind . . . is embedded in the Old Norse kenning "wind of the giantess"' (p. 18).

59 See Thomas A. Carnicelli, ed., *King Alfred's Version of Augustine's Soliloquies* (Cambridge, MA, 1969); see p. 32 for a list of the additions.

60 Carnicelli, ed., *Soliloquies*, p. 62, line 9. He thereby transforms Augustine's reference to the senses as the ship, whereby information is brought to the mind (land) where it is processed into the concept of the eyes as a ship, transmitting information to the mind, and in turn the 'eyes of the mind' as the anchor-cable securing the person (ship) to God. See Frank, 'Old English *ancor* "anchor"', pp. 16–17, on Alfred's introduction of the anchor-cable as a means of securing the ship of the mind. On Alfred's use of the anchor metaphor, see Ruth Waterhouse, 'Tone in Alfred's Version of Augustine's *Soliloquies*', in *Studies in Earlier Old English Prose*, ed. P. Szarmach (Albany, 1986), pp. 47–85, at pp. 64–8; and Milton McC. Gatch, 'King Alfred's Version of Augustine's *Soliloquies*: Some Suggestions on its Rationale and Unity', in the same collection, pp. 17–45, at p. 31.

Alfred's establishment of a royal navy, probably in 896,[61] it is reasonable to assume that ships were on his mind, so possibly practical considerations have given rise here to figurative language choices.[62]

While sea travel has classical and patristic precedents as a conventional motif for the imagination, the high number of journeys recounted and the prevalence of nautical metaphors for imagination, especially in the elegies of the Exeter Book, suggest the full absorption or creation of these concepts in the vernacular mind.[63] A recognisable feature of these lament poems is their metaphoric construction of mental anguish in terms of the physical realities of the maritime world, the mind as a ship, and thought as sea travel. Given the harsh realities of the seafaring life in north-western Europe in the early Middle Ages, and the convention of configuring the sea as a destructive, combative force to be dealt with by travellers, it is not surprising to see maritime metaphors being deployed in order to express difficulty and despair to be reckoned with, and in turn potentially, the means of escape from worldly constraints.

In Old English poetry, both the fictional narrative context and the metaphoric language are nautical. Here, a variation on this ship of the mind metaphor develops, where the mind is not only conceived of as a ship, but also, its imaginative capacity is activated by sea-travel. This schema, of the mind as a wandering entity, seems to figure within a broader cultural fascination with travel: the narrative preoccupations of our extant Old English texts, particularly but not only the poetry, suggest that the literate Anglo-Saxon was interested in hearing about voyages, both actual and imaginary, to remote locations. The actual sea-voyages of, among others, Andreas to Mermedonia, Elene to Jerusalem and Beowulf to and from Denmark, come to mind, as well as other journeys recounted in the Old English translations of Orosius's and Bede's histories, and referred to in *Widsith* and *Deor*. More broadly, 'The numerous references to sea and island in Bede and Boniface, for example, are no empty formulas but necessary images for remembering the past'.[64] Similarly, there is some evidence of both merchant and pilgrim sea travel from England to Italy until at least the late ninth century.[65] Michael McCormack's detailed study of early-medieval maritime activity has provided

61 Eric John, *Reassessing Anglo-Saxon England* (Manchester and New York, 1996), p. 83.

62 Frank, 'Old English *ancor* "anchor"', notes that Alfred accounts for half of the extant occurrences of 'anchor' in Old English (p. 15). On the prevalence of the metaphors of 'ship of the mind' and 'eye of the mind' in Alfredian translations as evidence of common authorship and Alfred's developing epistemology, see Miranda Wilcox, 'Alfred's Epistemological Metaphors: *eagan modes* and *scip modes*', *Anglo-Saxon England* 35 (2006), 179–217.

63 The most comprehensive study of this genre is A. L. Klinck, *The Old English Elegies: A Critical Edition and Genre Study* (Montreal and Kingston, 1992), pp. 223–51. See also: S. B. Greenfield, 'The Old English Elegies', in *Continuations and Beginnings: Studies in Old English Literature*, ed. E. G. Stanley (London, 1966), pp. 142–75; and P. E. Head, *Representation and Design: Tracing a Hermeneutics of Old English Poetry* (New York, 1997), pp. 120–1. Many critics prefer the idea of the elegiac mode or mood, as expressed in B. J. Timmer, 'The Elegiac Mood in Old English Poetry', *English Studies* 24 (1942), 33–44.

64 Nicholas Howe, *Migration and Mythmaking in Anglo-Saxon England* (New Haven and London, 1989), p. 34.

65 McCormack, *Origins of the European Economy*, p. 679.

corroboration of some of the details provided in Old English nautical poetry, that medieval ships were often 'small and maneuverable',[66] that night travel and even night departures were common,[67] and travel during the winter months not especially rare, made possible by the practice of sailing small vessels close to the shore where shelter was readily available.[68] Such practices explain the rationale behind the commonplace configuration of ships as horses or as sentient beings involved in the contest against the elements in Old English (and Old Norse) poetry: the ship in *Beowulf*, for example, is presented as a spirited participant in or projection of the hero's magnificent journeys: it is described as exhilarated (line 220) and resting peacefully (line 301), awaiting its owner and master (line 1882).[69]

The mind as a wandering entity: The Wanderer and The Seafarer

However, in *The Wanderer* and *The Seafarer*, such common 'ocean steed' metaphors are absent: the ship is used as the vehicle of the metaphor rather than being the object itself of metaphoric expression. Yet these two poems do reflect a local predilection for journeying, and the cultural centrality of the sea to Anglo-Saxon life, in that when acts of the imagination are recounted, metaphoric constructions of the mind and its activities tend to be based on physical sea travel, and to rely on the figurative potential of the ship. These verse texts describe the harsh realities of actual maritime wandering as well as the two-edged aspect of mental travel expressed through seafaring metaphors. Critics have usually followed the options established by early editors of these poems, in understanding their voyages as either Christian metaphors of exile, accounts of actual journeys, or as conventional elegiac motifs.[70] But in these two poems, the maritime context specifically enables both the imagined act of cognition, and its literary representation. The mind itself is constructed as a secret enclosure to be controlled, but also as a travelling entity, able to move both metaphysically and metaphorically: it seems to be able to travel outside the body somehow, and also to leave the present via remembrance, dreams, imagination and reverie. Through the deployment of dual metaphorical schemas, these poems attempt to conceptualise the conundrum of an embodied mind.

66 Ibid., p. 422.

67 Ibid., p. 426.

68 Ibid., p. 468.

69 Similarly, *Andreas* 438–49. On metaphors configuring ships as beasts, see Karin E. Olsen, 'Animated Ships in Old English and Old Norse Poetry', in *Animals and the Symbolic in Mediaeval Art and Literature*, ed. L. A. J. R. Houwen, Mediaevalia Groningana 20 (Groningen, 1997), pp. 53–66.

70 Ida Gordon, ed., *The Seafarer* (London, 1960; repr. Exeter, 1996), pp. 5–10, outlines early debates on the allegorical or realistic sea journey depicted in *The Seafarer*, opting herself for the elegiac motif argument (p. 21) in this text and *The Wanderer*. R. F. Leslie agrees in his edition of *The Wanderer* (Manchester, 1966), though favours native Germanic over Celtic origins to the motif (p. 36). T. P. Dunning and A. J. Bliss, eds, *The Wanderer* (London, 1969), likewise explain the sea setting in terms of poetic convention (p. 96). More recent criticism, though it has explored the diction, syntax and literary techniques of these poems, has not ventured beyond these possibilities for the maritime setting.

For Anglo-Saxon poets, this specifically containing mind determines individual character by virtue of its contents, which requirement management, but is imagined as being far from static; much of the extant poetry relies upon a schema of thought as figurative travel, which is likewise under personal control. The narrator of *Genesis B*, for instance, reports that Eve 'hire mod ongan / lætan æfter þam larum' (591b–592a), 'let her mind go along with those counsels', in a typical reference to thought as movement that is still idiomatic today. A corollary of this pervasive conception of thought in terms of physical movement is the evident literary interest in the predilection of the human mind to wander away from the here and now, giving rise to the expressed moral imperative to keep in check that tendency to stray, and to batten down the mental hatches securely.

Many extant Old English poems express an interest of this kind in the mental life, and exemplify the linguistic habit of mentioning the mind in contexts treating emotion, wisdom and personal mastery. This cultural interest in the life of the mind is particularly evident in verse that suggests its own rhetorical and didactic potential to encourage mental vigilance in an audience.[71] In the most extensive and effective examples, that cultural focus on the mind combines with another Anglo-Saxon preoccupation – the sea – to explore the compelling force of imagination and the metaphysical aspect of the embodied mind that requires both careful control and stability.

The Old English verse that deals most extensively with the peripatetic mind has a distinctly and pervasively maritime flavour, and often, the context in which the mind is imagined to wander, and the narrative opportunity for expressing the need for mental vigilance, arises from a physical journey across the sea. For instance, *The Wanderer* talks about sending the mind out over the sea (55b–7):

Cearo bið geniwad
þam þe sendan sceal swiþe geneahhe
ofer waþema gebind werigne sefan.

Care is renewed for the one who must very often send his weary mind over the shackle of the waves.

Similarly, the narrator of *The Seafarer* refers to the flight of the mind over the sea and back again (58–64a):

Forþon nu min hyge hweorfeð ofer hreþerlocan,
min modsefa mid mereflode
ofer hwæles eþel hweorfeð wide,
eorþan sceatas, cymeð eft to me
gifre ond grædig, gielleð anfloga,
hweteð on hwælweg hreþer unwearnum
ofer holma gelagu.

71 I have detailed this 'cultural focus on the mind' in *The Life of the Mind*, building on the line of inquiry opened up by Godden, 'Anglo-Saxons on the Mind'. See also my 'The Medieval Concept of the Self in Anglo-Saxon England', *Self and Identity* 1 (2002), 77–97; and 'Thinking in Metaphors: Figurative Language and Ideas on the Mind', *Sydney Studies in English* 30 (2004), 3–20.

> Therefore, now my mind turns beyond the confines of the chest, my mind widely turns amidst the sea-flood over the homeland of the whale, the expanses of the earth and comes back to me eager and greedy, the lone flier yells, incites the mind irresistibly on the whale-road over the expanses of the ocean.

Both these instances, from the most anthologised, admired and analysed short texts of the extant Anglo-Saxon corpus, explicitly deal with sea travel, and express imagination via conceptual metaphors derived from nautical activity. The conflation of the actual and the metaphorical journey, and the conventional associations of the sea, early morning, melancholy, and physical and mental weariness, are critical commonplaces in modern scholarship. Eric Stanley's work on 'Old English Poetic Diction' has long been a foundation for literary interpretations of the maritime content of poetic language,[72] deployed more recently in Peter Clemoes's analysis of poetic language dealing with the sea.[73] Clemoes's idea, that 'seafaring fascinated poets as a recurrent text, or demonstration, of human endurance and skill'[74] and that maritime diction encoded that physical and mental challenge, is convincing.[75] It reminds us of the symbolic resonances of the sea in this literary culture. It is certainly likely that seafaring comes up often in our extant Old English poetic corpus because it was 'symbolic in evoking both social and natural forces',[76] but this reason for privileging sea voyages in literary narratives does not fully account for the way context is coupled with content: the way in which sea journeys are apt to bring about remembrance, reflection and philosophical rumination. So, while critics routinely mention the variety of nautical terminology in Old English poetry, and the maritime flavour of the homiletic, patristic and Germanic motifs evident in *The Wanderer* and *The Seafarer* in particular,[77] those contexts are yet to be linked to Anglo-Saxon modes of cognition or representations of a thinking mind within a physical body.

Both texts feature the same two human activities of body and mind – travel and thought – though representation is complicated by the probably figurative nature of the language used to depict them and the possibility that the entire texts are allegorical. Notwithstanding attempts to relate the ideas and events in our poems to actual practices such as exile or seafaring, or motifs such as *peregrinatio*,[78] most readers would agree with Stanley's conclusion

72 Eric Stanley, 'Old English Poetic Diction and the Interpretation of *The Wanderer, The Seafarer,* and *The Penitent's Prayer*', *Anglia* 73 (1955), 413–66, at pp. 433–4. 'The early morning is a time of terror without solace', an example of the tendency whereby the Anglo-Saxon poets 'use the processes of nature as symbols of moods' (p. 452).

73 Peter Clemoes, *Interactions of Thought and Language in Old English Poetry* (Cambridge, 1995), pp. 134–50.

74 Ibid., p. 146.

75 In *Genesis B*, when Adam is attempting to appease God by expressing his own willingness to make amends, he expresses the ultimate personal endeavour: he offers to undertake a sea journey, should that be necessary (lines 828–35a).

76 Clemoes, *Interactions*, p. 147.

77 For example, 'The striking attachment of Anglo-Saxon poets to descriptions of voyages against a background of story seascapes', in *The Wanderer*, ed. R. F. Leslie, p. 36.

78 See Patrick Cook, '*Woriað þa Winsalo*: The Bonds of Exile in "The Wanderer"', *Neophilologus*

that 'the Anglo-Saxons were fully conversant with figurative thought',[79] as demonstrated throughout both *The Wanderer* and *The Seafarer*. It is now evident that, besides the commonplace metaphors of the soul as a fortress and the body as an enclosure for the mind, many metaphors of thought and imagination are used in these texts, though the connotations and referents of those expressions have not yet been fully explored.

A more specifically cognitive approach allows a more detailed consideration of Old English conceptual metaphors to inform the developing picture of vernacular mind-schemas and their literary expression. The varied and multiple figurative representations of the mental life in Old English texts, and their parallel existences in both analogues and in unrelated texts, speaks against a single prevailing conceptual metaphor for the mind in this culture perceived by some scholars (i.e. the container model). Rather, the evocative language of poetry articulates, but leaves unresolved, the complexities of consciousness – the view from within, a conundrum that Cognitive Psychology and Neuroscience are still struggling to explain – in multiple and competing ways. Old English poetry articulates these complexities through the combination of the universal mind-as-container metaphor with the more particularly vernacular idea of the mind as an entity wandering in a maritime context. Such metaphoric clustering achieves a conceptual blend that expresses the paradox of a metaphysical mind within a physical body.

The challenges of directing the restless mind and the fluidity of its thoughts are mapped onto the perilous and onerous maritime voyage. In *The Wanderer*, when remembrance produces sorrow (50b–1), the misery of the present is constructed out of a dislocation between the mind and the self expressed as a personal challenge: to contain thoughts in the mind and the mind in the body (55b–7, cited above, p. 41). Here, perhaps by analogy with the idea that thoughts travel, the mind itself is constructed metaphorically as a maritime vehicle, bird, or beast of burden, which can be compelled by the self ('sendan sceal') to travel through time and space. When the speaker dreams and dreams again, he reignites sorrow, 'þonne maga gemynd mod geondhweorfeð' (51), 'when the memory of kinsmen passes through the mind.'[80] In this metaphoric expression, the mind is configured as a room, and memory as a travelling entity over which the speaker, in terms of his core being, or self, has only incomplete control. The moral imperative to restrain the wilful mind, and the act of will necessary to remain reticent, is articulated in *The Wanderer* via a metaphor of mind as a fortified keep for imprisoning frenetic thought (11a–14):

80 (1996), 127–37; and Dorothy Whitelock, 'The Interpretation of 'The Seafarer', in *The Early Cultures of North-West Europe*, ed. Cyril Fox and Bruce Dickins (Cambridge, 1950), pp. 261–72.

79 Stanley, 'Old English Poetic Diction', p. 452.

80 This sentence can and has been construed the other way round: 'Sorrow is renewed when the mind passes through the memory of kinsmen'. It is widely recognised that both readings are syntactically possible and contextually appropriate. See B. Mitchell and F. C. Robinson, *A Guide to Old English*, 5th edn (Oxford, 1992), p. 272, though the version provided above in the text seems preferable contextually.

Ic to soþe wat
þæt biþ in eorle indryhten þeaw,
þæt he his ferðlocan fæste binde,
heald his hordcofan, hycge swa he wille.

I know as a truth that it is an excellent virtue in a man that he firmly bind up his chest, hold fast his treasure-chamber, think what he will.

This obligation to restrain the mind is articulated elsewhere in the extant corpus, most notably in the very similar lines near the opening of *Homiletic Fragment II*: 'heald hordlocan, hyge fæste bind, / mid modsefan' (3–4a), 'hold fast the treasure chest, firmly bind (your) thought within the mind'.[81] In an even more explicit use of the metaphor of the will as the shackles restraining thoughts or even the mind as prisoners within the chest, *The Wanderer* urges the increased necessity for imprisoning the travel-prone mind during times of emotional distress (17–21):

Forðon domgeorne dreorigne oft
in hyra breostcofan bindað fæste;
swa ic modsefan minne sceolde,
. . . feterum sælan.

Therefore, righteous ones often bind firmly their sorrowful (mind) in the breast-box; as I must bind my mind . . . with fetters.

These repeated references to the metaphor of thought as movement out of an enclosure are particularly emphatic within those texts whose narrative relies on physical travel, *The Wanderer* and *The Seafarer*, where the restraint of the mind within the chest finds its symbolic counterpart in the containing, travelling qualities of the ship on the sea. The sequence of thoughts expressed in both these poems suggests, paradoxically, that compulsive mental activity has been at once transmuted into physical sea journeys, and also galvanised to a greater intensity by that same nautical travel. Maritime wandering is both an ostensible substitute for, and an actual outcome of, intense reflection on the past and rumination on the inner life, just as the imperative for absolute reticence is expressed, ironically, at great verbal length and elegance in these texts.

In *The Seafarer*, the expansive ocean becomes the metaphoric equivalent of the capacious mind, and wandering over the sea is the result of the mind's compulsion (33b–5):

. . . cnyssað nu
heortan geþohtas, þæt ic hean streamas,
sealtyþa gelac sylf cunnige.

Now thoughts press upon the heart, that I myself should explore the high seas, the tumult of the salt-sea waves.

81 Like *The Wanderer* and *The Seafarer*, *Homiletic Fragment II* contextualises the need for mind-restraint via the metaphor of troubles as the storms of life ('swa þeos woruld fareð,/ scurum scyndeð' (6b–7a), 'thus this world fares, impels with storms').

The mind both compels travel and itself travels over the sea. The ocean provides the context in which the mind's tendency for metaphysical travel becomes most intense: the mind reminds the *ferhð* to travel, but the *modsefa* itself turns about widely, suggesting almost an out-of-body experience for animated thoughts, activated by the physical circumstances of the embodied mind (58–64a, cited above, p. 41). The mind is frenetic and restless, bent on travelling widely alone. Like the setting of *The Wanderer*, here the ocean – encoded in multiple kennings – provides a suitably huge domain for this metaphoric as well as physical travel. As in *The Wanderer*, the familiar context of physical misery – encoded in the imagery of cold, confinement and solitude, most acute in the early morning – contrasts with the energetic, motivated and unconstrained mind. Just as the indicators of a sad mood have become symbolic, so too have the physical circumstances of the mind in the body provided a metaphorical schema for conceptualising the activities of imagination and memory.

Critics have debated the meaning of the reference to the apparent flight of the mind, or more often the soul, in *The Seafarer* and *The Wanderer*. F. N. M. Diekstra considers the reference to be to the metaphorical flight of the soul during contemplation, specifically 'an additional peregrinatio metaphor, supplementing that of the sea journey'.[82] Vivian Salmon believes *hyge* refers to the soul able to take on the form of a bird and thereby capable of escaping the breast to travel literally, not metaphorically,[83] a proposition which P. R. Orton endorses, with the qualification that the mind is specifically transformed into a cuckoo.[84] Similarly, Neil Hultin argues that in Anglo-Saxon England there existed a belief in the temporary flight of the soul from the body during life, as referred to in these poems.[85] Clemoes's idea that the poet reiterates Alcuin's notion of the mind's ability to travel beyond the body is, however, more credible,[86] suggesting the transfer of the concept, along with the metaphor, from a Latin to a vernacular context. Certainly, the conventional metaphor of thoughts as birds in classical and patristic texts,[87] as well as the related motif of imagination and dreaming as travel in both Latin and vernacular literature,[88] would have provided an ideal (and orthodox) conceptual framework for this metaphor of the mind itself as a bird.[89] If this interpretation is to

82 F. N. M. Diekstra, '*The Seafarer* 58–66a: The Flight of the Exiled Soul to its Fatherland', *Neophilologus* 55 (1971), 433–46, at p. 433.

83 V. Salmon, '"The Wanderer", "The Seafarer" and the Old English Conception of the Soul', *Modern Language Review* 55 (1960), 1–10.

84 P. R. Orton, '*The Seafarer* 58–62a', *Neophilologus* 66 (1982), 450–9.

85 N. Hultin, 'The External Soul in "The Seafarer" and "The Wanderer"', *Folklore* 88 (1977), 39–45.

86 Peter Clemoes, '*Mens absentia cogitans*', in *Medieval Literature and Civilization: Studies in Memory of G. N. Garmonsway*, ed. D. A. Pearsall and R. A. Waldron (London,1969), pp. 62–77, at p. 67.

87 M. Carruthers, *The Book of Memory: A Study of Memory in Medieval Culture* (Cambridge, 1990), p. 37.

88 Including Lactantius, Ambrose, Alcuin and Alfred's translation of Boethius. Godden, 'Anglo-Saxons on the Mind', p. 277.

89 The ascent of the soul to God, expressed as the mind with wings, occurs in the Old English version of the *Metres of Boethius*, most elaborately in IV, met. 1 (no. 24 in The Anglo-Saxon Poetic Records, ed. G. P. Krapp and E. van K. Dobbie (New York, 1931–42), V). See the new

be understood, perhaps we have here an example of a vernacular adaptation of a standard trope into a fully articulated image of a disembodied mind, in which independent travel is the foregrounded characteristic.

This specific discourse context, especially the account of an apparently physical journey during which the mental one occurs, provides a further dimension of meaning to this metaphoric language. The 'journey within a journey' idea which these poems both express not only reflects the complex dual speech situation most readers understand, but also promotes an image of life as a journey both of mind and of body. The didactic thrust of these texts relies on the culturally encoded convention that the sea can symbolise the imagination and also provide the ideal context for its activation: both are boundless and can be traversed in an infinite number of ways. Interestingly, the journeys related in both *The Wanderer* and *The Seafarer* are more process- than goal-oriented, focusing on travelling itself, and the hardships of the journey, rather than on reaching a destination. But this orientation changes within the conclusions of each poem, when the narrators look beyond the temporal world to the eternal one, by means of invoking a new set of travel metaphors.

Both poems conclude with a reference to the heavenly home, the final Christian destination. *The Seafarer* ends with the prayer that commences: 'uton we hycgan hwær we ham agen, / ond þonne geþencan hu we þider cumen' (117–18), 'let us consider where we have a home and then ponder how we may come there'.[90] *The Wanderer* more obliquely finishes with: 'wel bið þam þe him are seceð, / frofre to fæder on heofonum, þær us eal seo fæstnung stondeð' (114b–115), 'well is it for the one who seeks grace, the comfort of the father in heaven, where for us all stability stands'. Both poems rely for their homiletic peroration on the familiar allegory of life as a journey,[91] at the end of which heaven is home, as well as the implied need to contemplate this final destination.

The presence of these endings in poems, which feature sea-travel and metaphoric travel so prominently, expresses the need both to conceptualise life as a journey and to appreciate the salvific product of perceiving the resemblances between the actual and the figurative, decoding the metaphor, and constructing perception and belief by means of the systems invoked. These didactic exhortations valorise the central trope of rumination as a wide-ranging and energetic activity,[92] conflating actual and metaphoric travel into a

edition: Malcolm Godden and Susan Irvine, eds, *The Old English Boethius: An Edition of the Old English Versions of Boethius' De Consolatione Philosophiae* (Oxford, 2009).

90 A commonplace of Christian literature, including in texts produced in Anglo-Saxon England. See for example the opening of *Guthlac A*, where an angel welcomes to heaven a soul 'tidfara', 'a traveler the time of whose journey has come' (BT, s.v.): 'Nu þu most feran þider þu fundadest/ longe ond gelome' (6–7a), 'Now you must travel whither you have striven long and often'. *Resignation A* is also thematically oriented on the idea of life as a journey, specifically the metaphor of worldly existence as exile on the cold, harsh sea (59–64a). *Exodus* likewise evokes the heavenly reward after the dangerous temporal journey (1–7).

91 As argued in Irvine, *The Making of Textual Culture*, during the Middle Ages, '*allegoria* was understood to be a species of metaphor', p. 105.

92 Cf., the metaphor of the clean beast chewing its cud, or ruminating, for the act of poetic

powerful Christian statement of the importance of destination and the erratic and uncertain nature of travel in any direction other than the heavenly one.[93] This particular metaphoric expression, of life as a journey, thereby creates a self-reflexive argument for the importance of poetry like itself as a textual inspiration to meditation, and implies further metaphors – text as external memory, and memory as a tool for self-reform – both ideas developed by St Augustine of Hippo[94] and transmitted to Anglo-Saxon England with the importation of Christian learning.

In *The Seafarer* this imported Christian imperative to travel purposefully rather than aimlessly is encoded in the vernacular version of the maritime metaphor of steering the vessel of the mind: 'Stieran mon sceal strongum mode   ond þæt on staþelum healdan' (109), 'A man must govern an unruly mind and keep it in (its) place'. The metaphor of the mind as a ship on the rough seas which needs anchoring by faith is found in classical and patristic sources, notably Gregory's *Pastoral Care*, and is found widely in Old English literature,[95] including, in almost identical words, in *Maxims I*: 'Styran sceal mon strongum mode' (50a), 'A man must govern an unruly mind'. The theme is elaborated on in this poem, with nine more lines (51–9) that seem to be about the perils of sea-travel, but, given their proximity to line 50 about mental control, work effectively as an implied metaphor for restraint in the stormy waters of life, a concept the poet makes more explicit at one point (54–6):

> Swa biþ sæ smilte,
> þonne hy wind ne weceð;
> swa beoð þeoda geþwære,   þonne hy geþingad habbað.

> Just as the sea is calm when the wind does not agitate it, so peoples are harmonious when they have come to an agreement.

In both *The Seafarer* and *Maxims I*, we can see the importation and local variation of the concept (that self-control brings order, peacefulness and concord) along with the metaphor (that agreement is harmony or a calm sea). In these poems, as well as *The Wanderer*, the conflation of the material and the imaginary is facilitated by the deployment of very familiar tropes – life is a sea voyage; thought is travel – within accounts of apparently actual travel. This habit of conflation allows one to see the multiple and culturally

creation, in the account of the poet Cædmon in Bede's *Ecclesiastical History* (ed. Bertram Colgrave and R. A. B. Mynors [Oxford, 1969, repr. 1992], bk. 4, ch. 24), a metaphor retained in the Old English translation, and used often in patristic literature. See Stock, *Augustine the Reader*, p. 219 and n. 64; and Sharon M. Rowley, *The Old English Version of Bede's Historia Ecclesiastica*, Anglo-Saxon Studies 16 (Cambridge, 2011), pp. 169–72.

93 There are many precedents in classical and medieval Latin literature for the active but selective use of the memory. See J. Coleman, *Ancient and Medieval Memories: Studies in the Reconstruction of the Past* (Cambridge, 1992), p. 127; and more generally, on the metaphor of mediation as travel throughout the memory. See Carruthers, *The Craft of Thought*, p. 115.

94 On Augustine of Hippo's theories of the self-reforming aspects of active reading, see Stock, *Augustine the Reader*, esp. pp. 243–78. For the role of memory in this process, see Carruthers, *The Book of Memory*, pp. 156–88.

95 On this metaphor, see Frank, 'Old English *ancor* "anchor"', esp. pp. 17–18; and Waterhouse, 'Tone in Alfred's Version of Augustine's *Soliloquies*', pp. 64–8.

loaded literal and figurative meanings of lines such as the opening of *The Seafarer*: 'bitre breostceare gebiden hæbbe, / gecunnad in ceole carselda fela' (4–5), 'I have experienced bitter cares of the heart, explored in a ship many abodes of sorrow'.

In *The Wanderer* and *The Seafarer*, the power of the conventional metaphor of life as a journey combines with the familiar themes of the life of exile as lonely, cold and burdened with seafaring, to provide a vivid, if conventional, backdrop for the central focus on the frenetic, uncontrollable imagination, itself a maritime wanderer. This combination of actual with metaphoric journeying blurs the line between the figurative and the literal in these verse texts and highlights the cultural knowledge required to interpret such systematic coding via conventional metaphor. The presence, even dominance, of the metaphoric maritime journey in these texts, coupled with the conventional status and symbolic function of such journeys in earlier Latin literature, suggests too that a considerable part of the poems' textual meaning is encoded in their use of two privileged metaphors for representing embodied human consciousness.

Beyond the idea that life was imagined as a voyage, the ideas of mind and sea were coupled more widely in this literary culture, suggesting the existence of an overarching conceptual metaphor linking the two. For instance, in *Wulf and Eadwacer*, the absoluteness of the separation of the lovers is expressed through the image of the uncrossable sea: 'Wulf is on iege, ic on oþerre' (4), 'Wulf is on one island, I on another'. Similarly, in *The Wife's Lament*, the dear lord is imagined living far away in a desolate place, beside the sea, prevented from making a journey (46–51a). In *Andreas*, God is disguised as a sailor, in a sustained allegory of life on the rough sea, recounted during an actual ocean crossing (511–36), after which Andrew and his men are transported to Mermedonia in a dream. Similarly, at the end of *Christ II*, the holy steersman provides safe harbour (850–66). A safe conclusion to a dangerous voyage is one example of the use of the concrete world to represent the immaterial, 'the way the mundane asserts itself as part of a Christian system of symbols'.[96] But the degree to which the sea is embedded in this set of associations is evident on the linguistic level, and in the proximate pairings of the mental and watery realms. The inter-association of mind and sea is evident in poetic collocations and juxtapositions, such as: 'merewerges mod' (*Seafarer* 12a), 'the sea-weary mind'; Andreas says he knows nothing of the Mermedonians mentality, nor of the sea highways (198b–201), and is reproached by god for hesitancy about the sea journey and feebleness of mind (211–12a). Similarly, the troubled mind contemplating sea travel is recounted in *Resignation B* (96b–99):

Ic bi me tylgust
secge þis sarspel ond ymb siþ spræce,
longunge fus, ond on lagu þence.

96 Patrick W. Connor, 'Religious Poetry', in *A Companion to Anglo-Saxon Literature*, ed. P. Pulsiano and E. Treharne (Oxford, 2001), pp. 251–67, at p. 258.

I narrate this sad story entirely about myself and speak about a journey, eager with longing, and think about the sea.

There is likewise a strong conflation of imagination and the sea journey in *The Husband's Message*, where the speaker urges his audience to make a sea trip, saying (8b–11a):

Eom nu her cumen
on ceolþele, ond nu cunnan scealt
hu þu ymb modlufan mines frean
on hyge hycge.

I have come here in a shipboard and now you must find out what you think in your mind about my lord's (mind-)affection.

Embodiment and cognition: integrated conceptual metaphors

All these instances suggest that Anglo-Saxon poets, and presumably their audiences, were not only familiar with the specific metaphors treated here, and used to the poetic exploration of the inner life, but had advanced facility in the nuanced use of conceptual metaphor more generally. From a cultural perspective, it is clear from these texts that the contents of the embodied mind were of great interest, as was the psychological impact of real or imagined physical hardship. There was a genuine understanding of the inherent interaction between the container and its contents, and a sustained interest in that relationship. More specifically, the mind and sea were conventionally paired, at least in poetic contexts, and evidence that the harsh conditions of maritime journeys provided a ready metaphoric reference for mental vigilance. Furthermore, the expansive dimensions and directional freedom of nautical wayfaring provided an appropriate figure for the freewheeling imagination, only partially confined by the human contrivance, the ship, itself representing the enclosing mind. Anglo-Saxon poets make extensive use of these habitual associations in order to express the paradox of the metaphysical mind in the physical body, at once both free and constrained. Any suggestion that Old English references to a corporeal mind were not necessarily metaphorical, but rather frequently reflected a common-sense view of a physical mind,[97] would be difficult to maintain in relation to the poetic examples treated here, especially given the widespread use of metaphor in Old English verse and the particularly resonant associations of the sea in this culture. Anglo-Saxon recipients of vernacular verse need not have been explicitly instructed in the 'Augustinian doctrine of the "incorporeal *anima*"'[98] in order to imagine the mind as anything other than a part of the body, and would have been assisted

97 A position argued by Lockett in 'Corporeality', who does ultimately qualify her argument to admit that, while she believes references to 'the hydraulic model' of the mind are usually literal, they can be metaphorical, and have to be interpreted individually (p. 367). The hydraulic model (and 'emergent metaphoricity') is presented again in the monograph developed from this thesis, *Anglo-Saxon Psychologies*.

98 Lockett, 'Corporeality', p. 283.

in their conceptualisation and understanding via the widespread figurative mapping of the attributes of the ship on a voyage onto the inner life.[99]

Even if the two dominant metaphors treated at length here – of the mind as container and as wandering entity – were logically irreconcilable, that would not undermine their rhetorical effectiveness: after all, we routinely use conflicting metaphors and can easily interpret a mixed metaphor. Furthermore, there is ample evidence to support the idea that literate Anglo-Saxons had a particular taste for paradox, enigmas, riddles and puzzles, so the paradoxical was clearly not only tolerated, but positively enjoyed, especially within poetic discourse contexts where the abstract, complex and metaphysical are the norm rather than the exception.[100] Indeed, the figure of the ship as a mind that both contains and travels provides at least a partial resolution of the paradox within the evocative language of poetry, which articulates but leaves unresolved the complexities of human consciousness.

Old English verse uses the world of ships, sea travel and the physical sensations of the maritime life to express the complexities of metaphysical matters, and to explore emotional experiences, through conventionalised and culturally specific frames for the mind and its activities. These nautical metaphors and references arise from a confluence of Latinate and vernacular imaginations, and are deployed to conceptualise and to articulate a human consciousness that is at once embodied yet metaphysical. In particular, the trope of the human mind as a ship on a voyage captures the paradox of a containing, wandering mind within a culture so mindful of the literal and metaphorical possibilities of the sea. This manner of organising the phenomenon of human existence reflects and sentimentalises a particular lifestyle and worldview that becomes a discursive construction of Old English poetry. The pervasive and flexible influence of the maritime imagination in Anglo-Saxon England provides a rich and pragmatic means of reflecting on the mysteries of the human mind, and conceptually underpins much of its most durable and appealing poetry. The particularly maritime imagination of the Anglo-Saxon poets shapes the conceptual frames and figurative language of the mind in this culture, in that the combined containing and travelling qualities of the ship enable the paradoxical conceptualisation of the mind both as a receptacle for thoughts and emotions and also as a travelling entity.

This analysis of the pervasive conceptual metaphors underpinning figurative expression and ideas about life and the human mind lends support to theories put forward within Cognitive Historicism. Specifically, it shows the culturally specific articulation of embodied experience in Anglo-Saxon poetry. This vernacular representation of the dynamic combination

99 James T. McIlwain, 'Brain and Mind in Anglo-Saxon Medicine', *Viator* 37 (2006), 103–12, presents evidence to support the contention that medical writings conceive of the brain as the executive organ of the body ('encephalocentric tradition'), whereas literary and religious texts favour the heart (the 'cardiocentric tradition'). McIlwain explains the existence of the incongruity by agreeing with my earlier published argument that *heorte* and *breost* were used metaphorically (p. 122).

100 Clemoes, *Interactions of Thought*, p. 103.

of immediate thought, memory of past moments, and hope for an eternal future is both socially and temporally situated, and also a product of the evolved neurobiological structures in the human brain that present-day readers share. Conceptual metaphors are enabled and determined both by the specific historical moment, and also by cognitive hard-wiring. As this chapter has suggested, then, a cognitive approach to the formation and transmission of language and concepts can thereby account for individual, as well as cross-cultural, intelligibility.[101] Widespread use of systems of metaphors betokens an evolved cognitive capacity for blending concepts to create new understanding, an important idea that is the subject of the next chapter.

101 An earlier version of portions of this chapter appeared in 'The Maritime Imagination and the Paradoxical Mind in Old English Poetry', *Anglo-Saxon England* 39 (2011), 21–42.

3

Conceptual Blending

The theory of conceptual blending

The creation and processing of metaphor is one instance of what has become known as 'conceptual blending'. This theory explains how the brain integrates information, and accounts for how the combination of ideas can be more than the sum of its parts. This theory is probably the most important concept to cross over from Cognitive Science to Literary Studies, both because of its capacity to explain central, complex issues, and also its broad applicability.[1] Blending theory accounts for how our brains are able to learn through the selection and combination of the old with the new: how structured schemas and background knowledge are integrated with emerging contextualised information in the formation of new concepts. It relies on the idea that sets of characteristics are mapped across into new contexts, where they are blended and integrated with other mappings, resulting in new meanings. In this way, conceptual blending shares structure with the closely allied conceptual metaphor theory, but goes beyond the latter's account of unidirectional mapping from source to target domain, to provide a way of understanding networks of cross-domain mappings in the construction of meaning. Like conceptual metaphor theory, this account of analogical conceptual projection is ideally suited to linguistic contexts, but applies more fundamentally to all human cognitive processing, and holds particular promise for theories of literary comprehension. In addition to accounting for figurative language use, it also provides a means of accounting for how counter-factual or hypothetical scenarios are created and integrated in information processing, and even more interestingly, allows structures of cultural knowledge to be identified in literary texts.

Since its emergence in the 1980s, conceptual blending (or conceptual integration) has developed as a framework for understanding the pervasive and dynamic mappings that characterise not only language use, but also human thought more broadly.[2] One of the most prolific and expansive scholars to write on this universal cognitive capacity is the literary specialist Mark Turner, who explains how conceptual blending is a basic and universal mental function whereby ideas are integrated or blended to create new concepts: 'The mental

1 F. Elizabeth Hart likewise puts cognitive blending at the top of the list of influential ideas to have crossed from Cognitive Science to Literary Studies: see 'The View of Where We've Been and Where We'd Like to Go', *College Literature* 33.1 (2006), 225–37, at p. 233.

2 As yet untested in experimental Psychology: see Peter Crisp, 'Between Extended Metaphor and Allegory: Is Blending Enough?', *Language and Literature* 17.4 (2008), 291–308, esp. p. 303.

operation of combining two mental packets of meaning – two schematic frames of knowledge or two scenarios, for example, – selectively and under constraints to create a third packet of meaning that has new, emergent meaning.'[3]

Turner has developed the blending framework in conjunction with the cognitive linguist Gilles Fauconnier,[4] who developed the Theory of Mental Spaces, a way of understanding the creation and mapping of conceptual structures for organising and processing discourse.[5] Mental spaces are interconnected, short-term belief structures that are built up from information in working memory in response to linguistic cues, and used and modified in the process of on-line meaning creation (unlike more permanent conceptual schemas that inform metaphorical mappings). Oakley and Coulson's definition of mental spaces is particularly helpful: 'Representations of the scenes and situations in a given discourse scenario as perceived, imagined, remembered or otherwise understood by the speaker. Mental spaces are used to package information about an interlocutor's center of interest within an interactive context.'[6] This cognitive semantic model of how meaning is constructed sets out convincingly the significant idea that cognitive constructions underpin language use,[7] which opened the way for even more fine-grained analyses and more sophisticated theoretical models of on-line meaning construction. The most appealing of these is Fauconnier and Turner's idea of blending, or 'conceptual integration networks', to account for the complex dynamics of how meaning is constructed in the mind through the creation of, and multiple mappings and connections across, mental spaces.

The idea of blending relies on an innate human ability to resolve disparate pieces of information and the conceptual scenarios they invoke (inputs) into new knowledge (a blend) that cannot be predicted from any of the constituent parts alone. Fauconnier and Turner refer to the system as a conceptual integration network in which there are four constituent parts or mental spaces: two inputs, one generic space (where the structure shared by the inputs is recruited and mapped), and one blended space, where partial structures from the inputs are synthesised into emergent meaning.[8] Using this basic schema and its variations, conceptual blending provides an account of how models stored in the memory are selectively deployed in networks of association

3 Mark Turner, 'The Cognitive Study of Art, Language, and Literature', *Poetics Today* 23.1 (2002), 9–20, at p. 10.

4 Mark Turner and Gilles Fauconnier, 'A Mechanism of Creativity', *Poetics Today* 20.3 (1999), 397–418; *The Way We Think: Conceptual Blending and the Mind's Hidden Complexities* (New York, 2002).

5 Gilles Fauconnier, *Mental Spaces: Aspects of Meaning Construction in Natural Language* (Cambridge, MA, 1985; repr. Cambridge, 1994), and *Mappings in Thought and Language* (Cambridge, 1997). See also Anders Hougaard and Todd Oakley, 'Mental Spaces and Discourse Analysis', in *Mental Spaces in Discourse and Interaction*, ed. Todd Oakley and Anders Hougaard (Amsterdam and Philadelphia, 2008), pp. 1–26.

6 Ted Oakley and Seana Coulson, 'Connecting the Dots: Mental Spaces and Metaphoric Language in Discourse', in *Mental Spaces*, ed. Oakley and Hougaard, pp. 27–50, at p. 29.

7 Eve Sweetser and Gilles Fauconnier, 'Cognitive Links and Domains: Basic Aspects of Mental Space Theory', in *Space, Worlds, and Grammar*, ed. Gilles Fauconnier and Eve Sweetser (Chicago and London, 1996), pp. 1–28, at p. 18.

8 Turner and Fauconnier, *The Way We Think*, p. 47.

with temporary models that are custom-built within an act of communication or interpretation.[9] In this theory of the dynamic organisation and synthesis of 'mental packets of meaning', the blend is imagined to occur within a mental space, or a temporary domain constructed for specific conceptual mapping. Individual metaphors operate in this way, as the most compact form of linguistic blends, and as surface manifestations of the fundamentally cross-mapping quality of the human conceptual system. A good illustration of this process at work is Grady and colleagues' example of the conventional meaning of the idiomatic metaphoric expression 'this surgeon is a butcher', in which incompetence is evoked, even though 'butcher' in isolation does not suggest this idea.[10] Grady and his collaborators emphasise that 'the notion of incompetence is not being projected from source to target',[11] but rather the blending of two mental spaces, one associated with surgery and one with butchery, produces a new emergent meaning: 'In the blended space, the means of BUTCHERY have been combined with the ends of, the individuals, and the surgical context of the SURGERY space. The incongruity of the butcher's means with the surgery's ends leads to the central inference that the [surgeon] is incompetent.'[12] The cognitive process of making meaning from such uses of language requires the mind to blend concepts and to make inferences beyond them, in response to their particular recruitment and combination, a process that cognitive scientists in many fields have been investigating for many years.

Literary scholars can both learn from and contribute to this developing research into mental operations, through the expansion and reapplication of highly nuanced tools for identifying and understanding the constitutive and constraining factors on contextualised linguistic meaning-making. As specialised readers, literary scholars are best placed to analyse how texts 'prompt for highly intricate blending'.[13] Blending theory is a particularly appropriate tool in this instance, in its focus on meaning construction rather than discourse processing per se, and its recognition of the emergent properties of meaning and their cognitive management. Further, conceptual blending holds even greater promise as a set of ideas that can be combined with other cognitive approaches treated in the chapters below to provide new paradigms for Literary Studies.

The idea of conceptual blending, like the pervasive, systematic nature of conceptual metaphor, is not new; that metaphor is fundamental to concept-formation was prefigured by Locke, Vico and others, and more recently discussed explicitly by Bühler and Stählin in the 1930s.[14] Fauconnier and Turner's theory, though, is different and innovative in its interest in the cognitive structure of

9 On this specific point, see Seana Coulson, *Semantic Leaps: Frame-Shifting and Conceptual Blending in Meaning Construction* (Cambridge, 2001), p. 202.

10 J. Grady, T. Oakley and S. Coulson, 'Blending and Metaphor', in *Metaphor in Cognitive Linguistics*, ed. R. Gibbs and G. Steen (Philadelphia, 1999), pp. 101–24, at p. 103.

11 Ibid.

12 Ibid., p. 106.

13 Turner and Fauconnier, *The Way We Think*, p. 414.

14 Brigitte Nerlich and David C. Clarke, 'Blending the Past and the Present: Conceptual and Linguistic Integration 1800–2000', in *Metaphor and Metonymy in Comparison and Contrast*, ed. René Dirven and Ralf Pörings (Berlin, 2003), pp. 555–93.

language and meaning, and has been developed energetically by many other scholars too, especially in Cognitive Linguistics.[15] It develops more directly out of Lakoff and Johnson's work on conceptual metaphor and their argument for the ubiquity of metaphoric thought and language, discussed above, p. 25. In its reliance on the idea of mental spaces in discourse, this theory of conceptual integration provides a cognitive linguistic way of modelling how people keep track of multiple, changing pieces of information, from which long-range as well as local meanings are synthesised. It is cognitive in that it attempts to analyse and describe the way in which human imagination processes and synthesises multiple pieces of information in the creation of ideas by using ideas and methods from Cognitive Science, particularly the idea that the mind functions within embodied experience.[16] It is linguistic in that it derives its data from human discourses, places metaphor in centre stage, and, in Turner's version of the framework, is based within the history of Classical Rhetoric.

Turner's work on cognitive rhetoric and metaphor in his books *Reading Minds* and *The Literary Mind*[17] underpins his deployment of the conceptual blending framework. In the latter book, he argues that the human mind is inherently metaphoric, and that the use of metaphor is not only pervasive and routine in human communication, but basic to human mental functions. Metaphor allows the human mind to operative imaginatively, and to map and blend the abstract onto the concrete, so conceptual metaphors such as UP IS GOOD reflect the bodily basis of being in the world subject to the force of gravity ('things are looking up', 'to move up in the world'), as discussed above, pp. 25–6. Turner argues that the human mind is naturally creative and able to synthesise and resolve complex combinations, in that its predisposition for story is the evolutionary precursor, rather than product of, language.[18] Here, Turner deploys ideas from Cognitive Neuroscience on perceptual image schemata to make a case for the biological basis of conceptual blends.[19] As a result of this innate mental ability, human beings are habitual blenders and possessors of an extensive capacity for the more intense conceptual integration required in processing literary texts. The evocative qualities and associative logic prompted by poetic language in particular rely on a high quotient of conceptual blending. Given the reliance of much poetry on highly charged or at least high-density or prominent imagery

15 Fauconnier and Sweetser, eds, *Space, Worlds, and Grammar*; Grady, Oakley, and Coulson, 'Blending and Metaphor'; Coulson, *Semantic Leaps*; B. Dancygier and E. Sweetser, *Mental Spaces in Grammar* (Cambridge, 2005); and the special issue of *Journal of Pragmatics* 37.10 (2005), on conceptual blending, edited by Seana Coulson and Todd Oakley. For a full list of items, see Mark Turner's website, at: http://markturner.org/blending.html

16 The initially tenuous link to Cognitive Science of Turner's original conceptual blending framework is explained in Tony E. Jackson, 'Issues and Problems in the Blending of Cognitive Science, Evolutionary Psychology, and Literary Studies', *Poetics Today* 23.1 (2002), 161–79, at p. 164. Edward Slingerland, *What Science Offers the Humanities: Integrating Body and Culture* (Cambridge, 2008), articulates the scientific support for the blended brain offered by Steven Pinker, Michael Arbib and Antonio Damasio, and the reliance on this capacity for the development of human language (pp. 180–1).

17 Mark Turner, *Reading Minds: The Study of English in the Age of Cognitive Science* (Princeton, 1991); and *The Literary Mind: The Origins of Thought and Language* (New York and Oxford, 1996).

18 Turner, *The Literary Mind*, p. 25.

19 Ibid., pp. 23–5.

and figurative language, this discourse type might be considered as the natural domain of the most concentrated blending. Indeed, the analysis of poetic imagery via Cognitive Semantics has shown just how fruitful this avenue into meaning-making can be.[20]

In Linguistics, Turner and Fauconnier's interpretive tool has been widely influential. In cognitive approaches to Pragmatics, for instance, especially in the consideration of cultural crossing, blending provides a key link in understanding the interplay of 'the meaning construction system (conceptual system) and the meaning prompting system (grammar)'.[21] This insight is particularly helpful in understanding how figurative language works. Because they provide a systematic way of accounting for how integration is triggered by language, Turner and Fauconnier's theory has been particularly embraced by scholars studying the way in which metaphoric language is apprehended and interpreted, and has been described as 'a psychological model of language and thought processes'.[22] But conceptual blending has attracted interest beyond Linguistics, especially its application to complex and highly contextualised discourse such as literary texts. Blends can be intricate as they can have more than two inputs, some of which might be less obvious than others, and some of which can be well established in the mind, and others new. Turner and Fauconnier use the phrase 'multiple blends' to refer to conceptual networks that feature multiple parallel or sequential inputs.

Literary texts routinely employ more elaborate blends in addition to metaphoric ones, not only symbols, images and even allegories, but also whole schemas from different concept domains, as in *The Dream of the Rood*, discussed below, pp. 59–63. Turner refers to talking animals in children's literature as one simple example, though points out that literary texts regularly place demands on readers to synthesise two or more whole schemas, scenarios or alternative realities, as in the case of inputs from the real world and imagined realities in a huge range of texts. Another of Turner's early examples is the figure of Satan in Milton's *Paradise Lost*: a blend of human and theological ways of being.[23] Literary texts, and other contexts that deploy figurative and evocative uses of language, imagery and imagined scenarios, tend to trigger the activity of blending in the mind of the textual recipient by virtue of their motivated use of language: Turner and Fauconnier insist that 'linguistic constructions prompt us to create conceptual blends',[24] an idea worthy of sustained consideration, especially when literary texts are at issue.

20 Line Brandt and Per Aage Brandt, 'Cognitive Poetics and Imagery', *European Journal of English Studies* 9.2 (2005), 117–30.

21 Istvan Kecskes, 'Editorial: Lexical Merging, Conceptual Blending, and Cultural Crossing', *Intercultural Pragmatics* 1.1 (2004), 1–26, at p. 4.

22 R. W. Gibbs Jr, 'Making Good Psychology out of Blending Theory', *Cognitive Linguistics* 11 (2000), 347–58.

23 Turner and Fauconnier, 'Mechanism of Creativity', p. 404.

24 Turner and Fauconnier, 'Mechanism of Creativity', p. 400. Scholars working in Cognitive Linguistics and Discourse Analysis have sought to explore 'how discourse functions as a vehicle of particular cognitive processes of conceptual integration' (Hougaard and Oakley, 'Introduction: Mental Spaces and Discourse Analysis', p. 13).

Blending as a literary technique

Taking Turner and Fauconnier's cue, many scholars have applied the ideas of conceptual blending and mental spaces to fictional narrative discourse, to such an extent that it has emerged as one of the major new directions for the study of narrative.[25] As a model for explaining the universal capacity of the evolved human mind to make mappings across concept domains, conceptual blending has become a hugely influential and useful concept in cognitive approaches to literature. For example, Elena Semino has demonstrated how Conceptual Blending Theory can be applied to the narrative fiction of Virginia Woolf.[26] Margaret Freeman shows how this theory makes poetic cognitive processing explicit and demonstrates the detailed and nuanced interpretation of poetry it makes possible in her analysis of the mapping strategies used by readers interpreting the poetry of Emily Dickinson and Sylvia Plath.[27] Freeman shows how blending theory provides a language with which to explain how coherence is created in the mind of the poetic recipient from apparently disparate pieces of information or images, and most usefully conceives of the poem itself as a single 'complex blend',[28] in contradistinction to Fauconnier and Turner, who use this phrase to refer to much smaller units of meaning. Freeman explains how the integration of mappings is constrained by the reader's motivation to synthesise clues to create the greatest coherence possible: 'a complex blend refers to the process by which multiple blends create "optimality crossovers" into each other's input spaces when "running" the blend'.[29] Blending theory aims to account for how this sense-making integration occurs at the conceptual level.

It is unsurprising that blending theory is particularly useful in the study of poetic meaning-making, given the reliance of verse on metaphor, whose defining characteristic is the blending of two concept domains into a single figurative expression. But this theory of conceptual blending applies more broadly to metaphoric qualities of thought and communication, as well as to expressions that technically qualify as metaphors, an idea of interest to scholars of narrative and literature more broadly. For instance, Barbara Dancygier's idea of 'narrative anchors', or flagged items of significance that act as 'place holders for mental spaces which cannot be fully set up or elaborated at a given point in the narrative',[30] is useful in its consideration of the way specific uses of language in texts participate in the creation of the larger narrative, and trigger actual or emergent cross-mappings and textual connections. These

25 Barbara Dancygier, 'The Text and the Story: Levels of Blending in Fictional Narratives', in *Mental Spaces*, ed. Oakley and Hougaard, pp. 51–78.

26 Elena Semino, 'Blending and Characters' Mental Functioning in Virginia Woolf's "Lappin and Lapinova"', *Language and Literature* 15.1 (2006), 55–72.

27 Margaret Freeman, 'Cognitive Mapping in Literary Analysis', *Style* 36.3 (2002), 466–85; and 'The Poem as Complex Blend: Conceptual Mappings of Metaphor in Sylvia Plath's "The Applicant"', *Language and Literature* 14.1 (2005), 25–44.

28 Freeman, 'The Poem as Complex Blend', pp. 28–9.

29 Ibid., p. 29.

30 Dancygier, 'Text and the Story', p. 58.

anchors determine the shape and content of the resolved narrative meaning, and rely on the specific network of blends triggered by a specific narrative.

The focus on emergent meaning and on-line meaning-construction made possible through a conceptual integration approach allows direct application to literary texts, an opportunity seized by a handful of scholars focusing on medieval texts,[31] though by no means yet exploited to its full potential. One area in particular in which blending theory could be applied fruitfully is in the consideration of cross-cultural mappings, in order to explore more fully how we are able to comprehend literary texts from remote cultures such as Anglo-Saxon England, and to consider if we make similar conceptual connections and blends to the contemporary recipients of these texts. Like the approaches detailed in the chapters that follow, Conceptual Integration Network Theory could supply a means of distinguishing the results of evolved cognitive processing from culturally influenced patterns of thought and schemas. Because it has been established that blending occurs subconsciously – 'conceptual integration typically works below the horizon of observation'[32] – and, like metaphoric language, is ubiquitous as a fundamental aspect of human cognitive functioning, we have here a way of examining textual traces or instantiations of culturally remote neurocognitive functioning in the complex, and motivated mappings that characterise the internal dynamics of literary texts.

Although blending has been shown to be used in many different ways in a variety of everyday and specialised communicative and imaginative contexts, it has been found to be particularly complex or prominent in our conceptual responses to literary texts. Indeed, because literary contexts trigger multiple, subtle, complex processes, and 'frequently prompt for highly intricate blending',[33] it is possible to count the degree of blending required by or prompted by a text as a feature of literariness. By this criterion, as by others, Old English texts such as *The Dream of the Rood*, which employs multiple complex blends, are highly 'literary'. The prominence of blended images and figurative language in Old English, and other poetry across times and cultures, suggests that it is not only conventional in poetic discourses, but indeed cognitively essential rather than culturally specific. In highly visual poetry, such as the dream vision, conceptual blending is not only a prominent cognitive mechanism deployed, but seems to determine an interpretive response chiefly dependent upon integration.

The following discussion of three distinct Old English texts demonstrates how poetic texts rely on conceptual blending, and how this cognitive framework for meaning-making is triggered by motivated language use, in concert

31 Craig Hamilton, 'Conceptual Integration in Christine de Pizan's *City of Ladies*', in *Cognitive Stylistics: Language and Cognition in Text Analysis*, ed. Elena Semino and Jonathan Culpeper (Amsterdam and Philadelphia, 2002), pp.1–22; and Peter Orton, 'Spouting Poetry: Cognitive Metaphor and Conceptual Blending in the Old Norse Myth of the Poetic Mead', in *Constructing Nations, Reconstructing Myth: Essays in Honour of T. A. Shippey*, ed. Andrew Wawn, Graham Johnson and John Walter (Turnhout, 2007), pp. 277–300.

32 Turner and Fauconnier, 'Mechanism of Creativity', p. 408.

33 Ibid., p. 414.

with intertextual, generic and cultural referents. The following discussion aims therefore to suggest how an analysis of Old English poetic texts can be enhanced by the ideas and methods of conceptual blending, through an exploration of the cognitive management of complex cross-mappings of ideas in these texts.

The Dream of the Rood – a complex conceptual blend

Mark Turner's analysis of literary texts provides a new way of interpreting figurative language that privileges what is going on in the mind of the recipient over other aspects of textuality, a perspective that makes particular reference to Old English literature, in that Turner has briefly applied his own theory of conceptual blending to *The Dream of the Rood.*[34] He treats, as 'a spectacular example of blending',[35] the multiple integration of the Cross with an articulated expression of its history, the Cross with Christ, the dreamer and the Cross, and the Cross and a thane, in this poem. In his listing of these 'many contributing scenarios',[36] Turner demonstrates the value of applying the cognitive lens to older texts, while himself stopping short of blending his own more broadly focused argument with either recent criticism on this poem or specific engagement with the language of the text. These lines of inquiry, then, are still open to specialists in the field. Despite the brevity and absolute quality of his remarks, Turner offers a timely reminder of what literary scholars, 'highly attuned to the intricate workings of creativity, invention, language, visual representation, and the construction of meaning'[37] have to offer Cognitive Science. He and others have opened the way for a more complete cognitive literary analysis of this and other Old English texts that specialist Anglo-Saxonists can and should take further.

One scholar from outside the field who has attempted something similar is Peter Stockwell, who conducts a more linguistically focused cognitive poetic analysis of *The Dream of Rood.*[38] His brief study of the poem entails a consideration of the narrative levels (and the requirement for interpretive transitions across those levels) of the dreamer recounting a vision of the Cross, which in turn recounts the events of the Crucifixion. He then moves on to a discussion of the world-schemas underpinning the text (i.e. the imagined reality where personified trees speak, which displaces the waking world of the dreamer), and the schematic knowledge required to interpret the text (i.e. the Gospel accounts of Christ's death). This analysis relies on the idea of 'conceptual upgrading', where the real world is transformed into the dream world (rather than the more usual reverse situation, where the dream world resembles more prosaic reality). Stockwell's conclusion, that 'the overall effect

34 Turner, 'Cognitive Study', pp. 14–15.
35 Ibid., p. 14.
36 Ibid.
37 Ibid., p. 18.
38 Peter Stockwell, *Cognitive Poetics: An Introduction* (London, 2002), pp. 82–7.

is to map transcendental concepts into the schema of the individual's personal sense',[39] accounts for the didactic and aesthetic effectiveness of this text, with a new emphasis on the interpretive process of the recipient. Like Turner's analysis, this one is relatively brief, and could potentially be extended by the Anglo-Saxonist to consider more fully the complex blending in this text. In particular, there is an opportunity to use the idea of conceptual blending to examine more closely the interpretive implications of the shifting and overlapping focalisation of the dreamer, Cross and Christ in the text; the selection, deployment and cross-cultural processing of vocabulary dealing with perception; and ideas centred on embodied universals and culturally specific concepts of memory. These ideas are addressed in turn below, pp. 61–3, following a brief account of the dynamic of the poem from a conceptual blending perspective.

The meaning complex that emerges from a reading of *The Dream of the Rood* is the result of the creation and integration of multiple metaphoric mappings. These mappings arise from the conceptual triggers created by the text, though also recruit culturally specific knowledge schemas from beyond it (e.g. from Scripture, other texts, the natural world and human social interaction). The poem has multiple input spaces, the two main ones being murder and redemption, with the Cross as instrument of both acts. Readerly expectations are set up and complicated by incongruous and counter-factual scenarios and pieces of information relating to the chameleon yet humanised qualities of the Cross and its divine context and salvific qualities. The cognitive management of all this information is directed by the drive for coherence, and constrained by the precise existence and arrangement of triggers in the text.

Freeman's idea of the whole poem as a 'complex blend' is useful here,[40] in that it can be considered to have some level of absolute internal coherence that results from multiple complex cross-space connections. In this poem, there are several blended spaces that operate within the dynamic megablend that is the poem: sin is presented as something that can be manifest on the body – blood, a physical burden, a blemish – and redemptive power is likewise visually manifest – as precious jewels. Similarly, the Christian Cross is presented as at once an instrument of torture and as a means of salvation; as both a person who can engage in a conversation, and an object; as a disempowered, maltreated figure and a source of enormous symbolic and real power. The reader is required to integrate these schemas in order to make sense of the text. As Freeman explains: 'Reading the poem involves running the blend. One input space provides the topological frame for the blend . . . . The other provides topological details which specify the values of the roles of the organizing frame.'[41]

In the case of *The Dream of the Rood*, the topological frame is provided by the blend of Crucifixion narrative; the details are provided by the dream

39 Ibid., p. 87.
40 Freeman, 'The Poem as Complex Blend', p. 29.
41 Ibid., p. 31.

narrative. Counter-factual details from the dream scenario (of changing coverings on the Cross, and its ability not only to talk but to experience feelings) are projected onto and blended with the frames provided by scriptural and exegetical narratives of the Crucifixion, images of the bejewelled tree of glory, and the conventions of the dream vision narrative. As a result, the images and ideas of the poem are organised to retain some structures from these spaces, and in turn to create a new megablend that is the poem. Meaning becomes dynamic when the arising blends become input spaces for further blends: for instance, when the crucifixion space becomes blended with the redemption space by means of the common factor of the Cross in its two functions, and when the addressee of the Cross's narrative becomes the speaker in a discourse situation with the textual recipient.

Individual metaphorical mappings are motivated by this overarching spine of logic in the text, and coherence is created by reference to it. At a technical level, the text prompts this process through two blends: the Cross with Christ, and the Cross with the dreamer. The narrator's metaphorical description of his emotional state, 'eall ic wæs mid sorgum gedrefed' (20b), 'I was completely afflicted with sorrows', is repeated a little later: 'sare ic wæs mid sorgum gedrefed' (59a), 'I was sorely afflicted with sorrows'. The blend more specifically recruits the mapping of wrongdoing onto physical blemish in the line 'ic synnum fah, / forwunded mid wommum' (13b–14), 'I was stained with sin, wounded with iniquities'. This structure is echoed in the Cross's remarks: 'eall ic wæs mid blode bestemed' (48b), 'I was completely drenched with blood', and 'eall ic wæs mid strælum forwundod' (62b), 'I was completely wounded with arrows', and its reference to its sorrow (80a). In the same way, the Cross's use of the term 'reordberend' (89b) recalls the dreamer's use of that same term at the opening of the poem (3b), and creates a category of 'voice-bearers', or other beings excluded from the current glorious visionary experience, as well as the blend 'speaking person in dream vision'. The similar use of blends involving THE EXTERNAL IS AN INDEX OF THE INTERNAL and THE PHYSICAL CROSS IS THE SOUL OF THE DREAMER provides by extension the further blends of CROSS IS DREAMER, and CROSS IS CHRIST as it recounts its suffering during the Crucifixion.

The reader or hearer of this text is required to manage the dynamic instability of the image of the Cross, by creating multiple overlapping sets of characteristics relating to jewels, blood and the human qualities of Christ, all of which provide characteristics of the Cross. In turn the mappings made from the Cross to the dreamer are mapped onto the recipient of the text, who is addressed after the end of the Cross's speech, from lines 122 to 156, but who is invited by the Cross to be one who 'in breostum bereð beacna selest' (118), 'bears in his heart the best of symbols'. The Cross and the dreamer act as intermediaries between Christ and the Reader, through the blending of human characteristics with symbolic qualities in the Cross.

This representation of the inner life through external features is repeated throughout the text, not only to create sustained symbolism, but also to establish a systematic connection between the inner life of the individual with the outer reality of the Cross, and a series of blends involving the two figures, and by analogy, with Christ. The reader is primed for this dual vision through the dreamer's account of the changing image of the Cross, from blood-spattered instrument of torture, to bejewelled symbol of hope: 'geseah ic þæt fuse beacen / wendan wædum and bleom' (21b–22a), 'I saw that noble sign change its coverings and colours'. Verbally, as well as symbolically, the Cross changes roles, doing duty both as representative of Christ and dreamer by means of blending. Like the dreamer, it is presented as one who sees, a beholder: 'ic þæt eall beheold' (58b), 'I beheld all that'. Similarly, the divine vision at the outset blends the Cross as the object of vision for both the saints and the dreamer (9b–12):[42]

> Beheoldon þær engel Dryhtnes ealle,
> fægere þurh forðgesceaft.   Ne wæs ðær huru fracodes gealga,
> ac hine þær beheoldon   halige gastas,
> men ofer moldan,   ond eall þeos mære gesceaft.

> All those beautiful by eternal decree beheld there the angel of the Lord – indeed that was not there the gallows of a wicked one – but the holy souls beheld it there, men over the earth, and all this glorious creation.

The scenario of someone observing a splendid vision is repeated throughout this text, resulting in a further aspect of the blend of the Cross with the dreamer.

Similarly, their roles in the discourse scenario unite them. The opening announcement of *The Dream of the Rood*, 'ic swefna cyst   secgan wylle' (1), 'I will recount the choicest of dreams', is repeated when the Cross, 'the best of woods' ('wudu selesta', 27b) recounts its own story. The use of the superlative in both instances allows a further cross-mapping of similarity and high quality between the narratives of the dreamer and the Cross, which in turn maps with the 'best of symbols' mentioned in line 118. The reader is predisposed to decode the shifting set of images recounted, and the conceptual blending among the dreamer, Cross and Christ, because mental processes typical of the dream experience are brought into play by means of such textual references to vision, remembered events and to dreams. Similarly, the capacity for words to create mental images is expressed in the instruction to the recipient: 'nu ic þe hate, . . . / þæt ðu þas gesyhðe   secge mannum, / onwreoh wordum' (95–7a), 'now I entreat you . . . that you recount this vision to others, reveal it with words'. Along with the opening remarks, and the emphasis on visual phenomena throughout the poem, these lines express Anglo-Saxon ideas about the ability of the human imagination to be stimulated by associative dream logic, expressed through the multivalent capacity of poetic discourse, to understand the complexities of faith. This literary technique relies on very particular ideas about imagination, textual interpretation and individual belief, concepts recoverable through this sort of analysis.

The interpretive implications of blended focalisation in *The Dream of the Rood* might include a fresh way of viewing the alignment provided by the

42 See also the discussion of this passage in Chapter 5 below.

verbal patterning developed through the descriptions of the Cross, Christ and the dreamer, in a system of cross-mappings where new attributes are permitted to emerge. For instance, the dreamer, like the Cross, is a witness to a visual phenomenon, and one who experiences emotions, then in turn recounts events. These roles are maintained within the different mental spaces created in the text, but blend as the referents of the chief narrative anchor, 'þas gesyhðe' (96a), 'this vision', are multiplied.

Anticipated reading practices are also embedded within the text, whose didactic power is reinforced by the omission of a waking scene, thereby allowing dream reality to spill over into lived reality implied in the narrative contract between narrator and addressee here. There is also a blend into eternal reality, in the final reference to the blessed life of the saved in the Christian afterlife, who 'wunedon on wuldre' (155a), 'dwelt in glory'. Here, the past and future of eternity are collapsed via the past-tense verb, 'wunedon', to refer to the saved already resident in heaven when Christ arrives, focalised by the Cross/Christ from the perspective of the arriver. These ideas act in concert with the clause 'þuhte me' (4a), 'it seemed to me', to frame the following account within a dream vision, but also to prompt the reader to activate and to blend together mental processes related to both dream experiences and to the interpretation of literary accounts of visions: the shifting and fragmentary representation of dream experiences, and the visual preoccupations of dream narratives.

An interpretation of this 'vision within a vision' requires the recipient to revise radically concepts from the real world, and to blend remembered dream experiences with known scriptural information and the personal experience of dream narratives. The reader's mind is thereby led to perceive a logical sequence from the narrative now, to a remembered vision, to the Cross's account of its origins and involvement in the Passion, and finally to the anticipated eternal reality of the heavenly afterlife. Because the mental processes anticipated by the creators of this literary text are the same as those modern readers use today, and textual cues are negotiated in the same way, cross-cultural interpretability is possible. Cognitive consistency across the ages permits a text such as *The Dream of the Rood* to engage the imagination and trigger complex blending operations of present-day audiences in much the same way in which it was consumed by contemporary ones.

Networks of blends in Riddle 43

A text as generically specific as a riddle deploys conceptual blends in slightly different ways from those evident in *The Dream of the Rood*, with different cognitive implications as a result of the generic reliance on the puzzle and an implied solution. But decoding a riddling text requires specific textual skill and contextual knowledge in order to overcome deceptive or incomplete uses of poetic language and obscure references, a task that becomes more difficult with temporal and social remoteness from the time of composition. It might therefore be useful to consider the riddle to be a selective road map of habitual associations, images and ways of thinking in double vision. Indeed, the figurative language employed in the Old English riddles perhaps hold special value as indicators of culturally specific blended concepts.[43] Their capacity for extended and complex ambiguity lends itself particularly to overlapping sets of ideas that can be analysed fruitfully as conceptual blends.

In *Riddle 43*, for which the solution is almost certainly 'soul and body',[44] this generic technique plays out via intersecting sets of binaries that set up a systematic network of blends. Here, the metaphysical/physical binary of the person is mirrored by the guest/host binary, as the latter pair stands in for, and suggests, the former in the riddling context. Similarly, the mortal and eternal life journeys are blended with those of guest schemas and customs of courtesy and social obligation implied in this text and in its cultural context.

Ic wat indryhtne æþelum deorne
giest in geardum, þam se grimma ne mæg
hungor sceððan ne se hata þurst,
yldo ne adle. Gif him arlice
esne þenað, se þe agan sceal
on þam siðfate, hy gesunde æt ham
findað witode him wiste ond blisse,
cnosles unrim, care, gif se esne
his hlaforde hyreð yfle,
frean on fore. Ne wile forht wesan
broþor oþrum; him þæt bam sceðeð,
þonne hy from bearme begen hweorfað
anre magan ellorfuse,
moddor ond sweostor. Mon, se þe wille,
cyþe cynewordum hu se cuma hatte,
eðþa se esne, þe ic her ymb sprice.

I know a stately, noble man, a guest within dwellings, whom fierce hunger, old age or disease cannot harm, nor heat make thirsty. If the slave, who must go on a journey, serves him honourably, safe at home they will find appointed to them food and happiness, but 'a countless progeny of sorrow if the servant obeys the lord, [his] master evilly on the way'.[45] One brother does not wish to be afraid of the other one;[46] that will harm them both when, eager to depart from life, they both leave the bosom of one woman, their mother and sister. The man, he who will, should declare in fitting words what the guest is called, or the servant, whom I speak about here.

[43] A point made elegantly by John D. Niles, *Old English Enigmatic Poems and the Play of the Texts* (Turnhout, 2006): 'through their bold use of metaphor, the riddles play with conceptual categories and habits of thought that were taken for granted in the culture of their time' (p. 12).

[44] See Mercedes Salvador, 'The Key to the Body: Unlocking Riddles 42–46', in *Naked Before God: Uncovering the Body in Anglo-Saxon England*, ed. B. C. Withers and J. Wilcox, Medieval European Studies 3 (Morgantown, 2005), pp. 60–96; *Riddle 43* is discussed on pp. 72–6.

[45] *DOE*, s. *cnôsl*, 1.a. figurative.

[46] *DOE*, s. *forht* 1.b. 'afraid, in awe (of something *dat.*): 'forht has alternatively been interpreted as "frightening": "nor should one brother wish to be a cause of fear to another"'.

This text, like many of the Exeter Book riddles, is difficult to translate smoothly into modern English, and is therefore open to competing interpretations. But we are able to make sense of it, and be fairly sure it refers to soul and body, because it evokes a number of recognisable yet culturally specific scripts (social norms, allowing inferential processes): not only of host and guest, lord and servant, but also of social hierarchy, the pleasures of the table, the inevitability of death and the threat of hunger, old age and disease as its most likely causes, of worldly prosperity, of fraternal loyalty and duty, the gendered associations of women with home and men with travel, and the privileging of honourable conduct and its rewards. In its discourse structure and address to the reader, it further invokes scripts of the value of the puzzle, and of elegant language, and thereby valorises its own existence self-referentially.

All these concepts are blended into a 'packet of meaning', in Turner's terms,[47] that calls upon the interpreter to supply the key link in the blend – the overarching metaphor and narrative anchor that guest is to soul and body is to host – in order for the whole system of meanings to hang together in a textually and logically coherent whole. The pieces of information provided are certainly selective, and constrained by generic and cultural relevance factors. The emergent meaning has a heuristic function of making vivid the reminder of life's transience through a number of blends associated with visiting and the home. The textual system of blends thereby urges the interpreter to supply the suggested supervenient blend in order for the solution of the riddle to be arrived at. This act of interpretation relies on the human capacity for this cognitive activity that readers today share with both the Anglo-Saxon composer and with contemporary recipients of the text.

In solving the riddle, the interpreter is invited to use the linguistic cues to trigger habitual and systematic conceptual associations – to deploy the complex conceptual system that is the product of embodiment and cultural situatedness, and given life in linguistic expression – to make the connection between the guest as a transient visitor in a house in which the gracious host is obliged to treat him well, and the soul, which is temporarily happy with the host, and from whom he must inevitably depart. It is remarkable that in a text which is almost a metaphor itself, and which relies on this central analogy, that there are no instances of individual conceptual metaphor, except perhaps 'mother earth'. But things are more complex than implied in the recognition of an overarching metaphoric relationship. There are two main blends operating here: between the lord/host and body and servant/guest and soul. In the social power relations mental space, attributes of lord are mapped onto body, and servant onto soul. In a second input space, the host and guest scenario, characteristics of the host are mapped onto the body, and guest onto the soul. There is a third blend, though, involving the family relationships mental space, where the soul and body are brothers, and life the earth. This distinction between the physical and the metaphysical is elided

[47] Turner, 'Cognitive Study', p. 10.

in the crossover that occurs within the poem to suggest that after death, the dual entity becomes singular, so a further blend of separated halves is implied.

This network of blends cued in by the text sets up a specific sort of interpretive process that is rare in the extant Anglo-Saxon riddles, very few of which could be described as spiritual allegory.[48] The didactic function of the text – to remind the reader of the transience of life and on the need to prepare for the hereafter – relies on the reader's ability not only to perform the conceptual blends implied by the text, but also to resolve the multiple blends evoked into a single 'megablend', or systematic pairing of blends, of an awareness that this life will end, and that the body should be the servant of the spirit, the host at the service of the guest.

The system of blends is clearly quite successfully implied, as this is one of the relatively few riddles for which there is a clear consensus of solution, and a discourse function aimed at exposing the spiritually marvellous (or existentially enigmatic) rather than baffling or titillating the reader as the other extant Old English riddles do. Its conceptual clarity and similarity with other texts invoking the soul and body relationship – including two poems of that name in the Exeter Book and the Junius manuscript – imply a pressing didactic desire to invite reflection on the fate of the soul through a consideration of its relationship with the mortal body. *Riddle 43* shows how a system of conceptual blends can perform a teaching function while conforming to generic and cultural norms, even in a relatively short text. In Turner's formulation, it deploys a 'conventional apparatus' of conceptual processing to express complex ideas in linguistic form.[49]

Blended perspectives in The Battle of Maldon

Blending theory can be usefully applied to more narrative texts, such as *The Battle of Maldon*, a poem that has attracted a huge amount of critical attention around the issues of historicity and rhetorical import. Critics have understood the text, probably composed shortly after the events of 991 it narrates, to express ideas relevant to specific historical and political context of late-tenth-century England, and also to celebrate admirable human qualities such as loyalty, bravery and resoluteness.[50] Its subject matter makes a central place for cognitive processing: the poem emphasises the mechanics of decision making, through its focus on the choices of the warriors on the battlefield, a thematic emphasis that has led critics to interpret the text as a piece of propaganda, or a 'mastering of contemporary problems by means of literature'.[51] Given

48 See Jennifer Neville, 'Fostering the Cuckoo: Exeter Book *Riddle 9*', *Review of English Studies* 58.236 (2007), 430–46, at p. 440.

49 Turner, *Reading Minds*, p. 66.

50 See Niles, *Old English Enigmatic Poems*, where the poem is described as 'a showpiece of contemporary ethics and politics' (p. 227); and Elizabeth Tyler, *Old English Poetics: The Aesthetics of the Familiar in Anglo-Saxon England* (York, 2006), pp. 160–1.

51 W. G. Busse and R. Holtei, '*The Battle of Maldon*: A Historical, Heroic, and Political Poem', *Neophilologus* 65 (1981), 614–21, at p. 619.

its interest in human choice-making, this poem is well suited to further cognitive-rhetorical analysis.

There is much that could be said on this subject in relation to this poem, but one instance that demonstrates the uses to which the conceptual blending approach could be put is Byrhtwold's famous speech at the end of the poem (309–19):

> Byrhtwold maþelode, bord hafenode,
> (se wæs eald geneat), æsc acwehte;
> he ful baldlice beornas lærde:
> 'Hige sceal þe heardra, heorte þe cenre,
> mod sceal þe mare, þe ure mægen lytlað.
> Her lið ure ealdor eal forheawen,
> god on greote. A mæg gnornian
> se ðe nu fram þis wigplegan wendan þenceð.
> Ic eom frod feores; fram ic ne wille,
> ac ic me be healfe minum hlaforde,
> be swa leofan men, licgan þence.'

> Byrhtwold spoke, raised his shield, he who was old brandished his spear; he instructed the men very boldly: 'The spirit must be the more resolute, the heart the keener, the mind must be greater, though our might diminishes. Here lies our lord, all hewn to pieces, the good man in the dirt. Ever may he regret it, he who intends to depart from this field of battle. I am old in years, but I will not go from here, as I intend to lie by the side of my lord, the dearest of men.'

This extract, like other Old English poetry, insists upon a reader's 'mental agility' and familiarity with the practice of a text being able to 'embody social thinking' and to present a 'complex vision of reality', as John D. Niles expresses it.[52] While Niles and other scholars rightly refer to this particular textual capacity for a social function and therefore the requirement for complex interpretation, engagement with the cognitive specifics of that act of interpretation has been avoided. Conceptual Blending Theory can address precisely this issue, and account for the network of ideas that pull in different directions in this poem.

To make sense of this passage, the interpreter must blend the context of the mead-hall boast and courtly pledges of allegiance – culturally determined bonds of loyalty and honour – with the rhetorical imperatives of the motivator of the losing side on a battlefield. More specifically, the reader must blend and reconcile the apparent incongruities of old age and verbal vigour; of moral strength within a weak position; of goodness and dirt; and, most extremely, to invert the conventional pairing in order to access the idea of regret with life, satisfaction with death. The code of military honour invoked in the rousing speech relies most on this final blend that evokes an unusual conceptual relationship, of the contract of loyalty with imagined posthumous respectability, and of personal glory in anticipated physical pain, destruction and

52 John D. Niles, *Old English Heroic Poetry and the Social Life of Texts* (Turnhout, 2007), pp. 1 and 236.

gore. In blending terminology, the 'personal benefits' input space is blended with the 'death' space by recruiting culturally entrenched views about nobility, loyalty and posthumous respectability. This resulting paradoxical blend of success in defeat, and strength in weakness, is the rhetorical lynchpin of Byrhtwold's argument, reinforced by the alliteration of the key, nominal, conceptually central words (e.g. 'mod', 'mare' and 'mægen' on line 313). It repackages physical defeat as metaphysical victory, and thereby implies the culturally valorised notion of reputation after death as a motivator of behaviour – at least in the rarefied world of heroic poetry – primarily by means of an inverted, and therefore novel, blend.

Here is a good example of the combined blend as more than the sum of its parts: death and glory are two prominent concepts in this literary culture, though the way in which they are blended in *The Battle of Maldon* creates a moving expression of the depths of loyalty and the extremes of sacrifice that can be imagined for the sake of honour. The addition of the repeated reference to the old age of the speaker do not make the sacrifice of any remaining life smaller, but rather adds to the poignancy of the passage through the blend of wisdom and experience usually associated with age in this culture, along with an abdication of age-related privileges. The elderly Byrhtwold demonstrates the extreme steadfastness of one who might otherwise be excused from making the ultimate sacrifice, and who must be the instrument of ensuring that his fellow fighters do not reach old age themselves. The interpreter blends these expressed and implied ideas in resolving the incongruities, conventions and surprises of this passage – and must try on a new conceptual system in order to make sense of the particular set of blends evoked here.

This process allows multiple perspectives to emerge, including that falseness, that 'significant and disturbing ambiguity', noted by Michael Swanton,[53] or the 'existence of paradoxes and wonders' recognised by Roberta Frank.[54] The interpreter takes a creative role in the integration of apparent anomalies. Blending theory can account for this interpretive role, and, further, possesses explanatory power for assertions such as that 'The *Maldon* poet has created a reassuring fantasy world in which what is truly at stake in battle is not treasure but a heroic subject formation and in which valuable objects taken from the battlefield carry with them the taint of shame.'[55] Indeed, it is the discomfort of making the blend – the uneasy incongruity of honour and shame in the same stretch of meaning-making – that is the source of the emotion in this textual encounter. The affective power of the text works only because the reader's cognitive and emotional energies are recruited in the act of textual interpretation via complex conceptual blending.

[53] Michael Swanton, '*The Battle of Maldon*: A Literary Caveat', *Journal of English and Germanic Philology* 67.3 (1968), 441–50, at p. 450.

[54] Roberta Frank, 'The Ideal of Men Dying with their Lord in *The Battle of Maldon*: Anachronism or *Nouvelle Vague*', in *People and Places in Northern Europe 500–1600: Essays in Honour of Peter Hayes Sawyer*, ed. Niels Lund and Ian Wood (Woodbridge, 1991), pp. 95–106, at p. 105.

[55] Bradley D. Ryner, 'Exchanging Battle: Subjective and Objective Conflicts in *The Battle of Maldon*', *English Studies* 87.3 (2006), 266–76, at p. 275.

Blending and meaning-making

Turner's theory of conceptual blending can illuminate the logistics of textual meaning-making in texts as diverse as *The Dream of the Rood*, *Riddle 43* and *The Battle of Maldon*. In the verse dream vision, as in the riddle and the historical narrative, multiple schemas are invoked and characteristics mapped across to new domain. The reader is thereby invited to blend them into a rich amalgam of emergent meanings in the act of sense-making. The poetic context supports this multivalence through evocative and information-rich metaphors, but also through syntactic juxtaposition, semantic ambiguity and the suppleness of vivid imagery, which behave as clues to linked conceptual systems that are highly culturally specific yet comprehensible beyond that context.

While there are clearly culturally determined domains of knowledge evoked in these examples (relating to scripture, dream vision, guest-host relations and acceptably loyal behaviour), there are recognisable processes at work in the conceptual projections that hold the texts together and allow readers beyond their contemporary cultural contexts to perceive coherence within them. These instances illustrate one way of approaching an analysis of the cognitive processes at work in our encounters with literary texts. But it is possible to construe Blending Theory as the basis for other cognitive theories, which account for the larger world schemas and fictional representations generated in response to textual cues and deployed within the act of interpretations, the subject of the following chapter.

4

Text World Theory

The theory of text worlds

Like the other cognitive approaches treated in the chapters above, Text World Theory provides a new way of understanding the mind's complex yet organised responses to any textual encounter, especially what occurs below the level of conscious awareness. This particular theory of discourse processing has developed out of Cognitive Grammar,[1] so has a linguistic origin, but has application beyond Linguistics because it looks well above the level of the sentence to the whole text. It assumes that to understand language we have to conceptualise its propositions; to create coherence from extended pieces of discourse, we have to keep track of those propositions in a systematic way, make inferences from them as a whole, and synthesise them with customised selections from stored knowledge. Text World Theory explains how this involved cognitive process works, by claiming that people make sense of discourse through the creation of mental representations of the ideas provoked by that discourse. These representations are not just images, but rather, they are conceptual systems that provide a way of managing and integrating collections of propositions. The mental representations created are seen within this framework as provisional worlds that must be imagined and then continuously updated in order to make sense of the text at hand, and to accommodate its emerging premises and assumptions. The process of creating coherence from these scenarios is cumulative and adaptive: the reader incrementally takes information from the text, with which general knowledge, stored schemas, inferencing and acts of the imagination are selectively combined within the reading process. The results include an evolving and custom-made discourse context and a collaborative act of meaning-making.

This activity of world-building is highly selective and flexible. When processing a piece of discourse, certain pieces of information are chosen and brought to bear in this concurrent (streamed or 'on-line') sense-making, a selective and synthetic cognitive process that can be analysed. For example, in trying to make sense of the lines, 'ðonne hit wæs renig weder   ond ic reotugu sæt, / þonne mec se beaducafa   bogum bilegde', 'When it was rainy weather

[1] For example, Ronald W. Langacker, *Cognitive Grammar: A Basic Introduction* (Oxford, 2008). The foundational principle of this grammar is the fundamental reliance of language on conceptualisation. This approach construes language as an inseparable component of cognitive functioning (p. 6), and grammar as symbolic of mental functioning (p. 4).

and I sat lamenting, then the battle-bold one embraced me with his arms',[2] a reader calls up memories, knowledge, associations with rain and stock situations that seem relevant, makes inferences, and imagines possible scenarios that are arranged and synthesised to allow all the clues to fit into some sort of coherent whole. The cognitive calibration of these pieces of information according to how well they fit with emerging coherence is complex, and is susceptible to detailed analysis that goes beyond available methods offered by close reading, reader-response, or other literary theories. When the following line is read, 'wæs me wyn to þon, wæs me hwæþre eac lað' (*Wulf and Eadwacer* 12), 'There was joy for me in that, yet there was also loathing', a process of repositioning idea-clusters and the recalibration of conceptual relationships occurs in the mind, whereby various possibilities and different emphases are entertained provisionally, until a way emerges of sensibly combining the ideas in response to creating a customised mental representation of the scene – a text world is created (perhaps involving a combination of sexual regret, bittersweet love and more generally an acknowledgement of emotional complexity, discussed further below).

This process is also collaborative. It is initially driven by the text, but developed further by the reader's memory, imagination and inference-making from textual cues. The text defines those areas of the knowledge-base immediately relevant to its meaning, to be supplied by the recipient. The specific collection of words on the page, along with the associations that those words evoked, prompt the reader to prioritise specific pieces of knowledge, relationship schemas and personal memories from the whole store of these ideas held in the mind, and to disregard other information; to make assumptions about information and beliefs shared with the textual producer; and to create certain expectations for further details (that are not in this case forthcoming, as discussed below). Text World Theory provides a way of analysing that process of selection, synthesis and image creation by focusing upon what is going on in the mind of the reader in response to the text on the page.

A text world is a conceptual space, or mental construct, created specifically from accumulated textual information, and built up from knowledge that the reader assumes is held in common with the writer. This assumed body of shared information is both crucial and mercurial: it is driven by the text, but ultimately governed by the reader's cognitive processing. This central idea of the text world as a representation of a motivated set of ideas relies on the mapping of physical experience onto conceptual functioning. Specifically, the experience of existing in the body and moving through physical space provides the working model for conceptual space. The text world framework, originally devised by Paul Werth, developed a means of theorising about how such conceptualised scenarios are custom-made in any linguistic encounter in order to organise and make sense of the specific

2 *Wulf and Eadwacer*, lines 10–11, discussed at length below, pp. 98–100. The textual difficulties of this sentence and the inscrutability of the poem in general are treated below, but are suspended here for the sake of the use of this citation as a textual sample only.

textual propositions provided.[3] Werth defines a text world as 'a state of affairs expressed in a discourse' that is modified by knowledge,[4] yet itself determines which pieces of knowledge are deployed. The text itself determines the parameters of relevance by virtue of both immediate contextual salience and ongoing continuity, which the reader relies on during the process of making inferences and deploying extra-textual knowledge. A text world may contain a number of 'sub-worlds' (or world changes or switches) distinguished by temporal or spatial shifts (e.g. change of scene, flashbacks), or movements into the minds of characters where hopes, fears and inclinations are registered. Humans construct these provisional models of the world evoked in a text in order to understand that world, a process possible only through the shared methods of cognitive functioning, embodied experience and information held in common by both textual creator and recipient.

Werth's Theory of Text Worlds is now used within Cognitive Poetics to account for the cognitive processing of discourse. Joanna Gavins in particular has developed and disseminated this framework for understanding how context-specific text-worlds, or mental models, are produced during the reading process, in her 'principled study of the conceptualization of language'.[5] This theory attempts to analyse the production and management of these imagined representations during discourse processing, and works well in conjunction with theories of conceptual blending, because it provides a means of accounting for the complex creation and synthesis of imaginative representations during reading. When we read poetry or fiction, we accommodate the existence of non-actual states of affairs and hypothetical situations. This process is possible only through the existence of referential frames, or 'whole chunks of experience and situations codified and stored in memory as single items' that are cued in by the text and brought to bear on sense-making.[6] Textual comprehension is thereby imagined as the exquisitely selective deployment of this frame knowledge to which new textual information is incrementally added, and which it in turn modifies: world-building whereby we 'alter mental representations to accommodate incoming information'.[7]

This literary-linguistic theory is closely related to the Theory of Mental Spaces in Cognitive Linguistics, developed by Gilles Fauconnier,[8] Elena Semino and their collaborators (also the basis of Conceptual Blending Theory discussed above, pp. 52–7),[9] but differs in its greater suitability for examining longer pieces

3 Paul Werth, *Text Worlds: Representing Textual Space in Discourse* (London, 1999). This book was produced posthumously by Mick Short.

4 Ibid., p. 206.

5 Joanna Gavins, *Text World Theory: An Introduction* (Edinburgh, 2006), p. 4. For a bibliography of recent and emerging work in the field, see Gavins' website: http://joanna-gavins.staff.shef.ac.uk.

6 Werth, *Text Worlds*, p. 20.

7 Gavins, *Text World Theory*, p. 10.

8 Gilles Fauconnier, *Mental Spaces: Aspects of Meaning Construction in Natural Language* (1985; repr. Cambridge, 1994); *Mappings in Thought and Language Language* (Cambridge, 1997). See also G. Fauconnier and E. Sweetser, eds, *Spaces, Worlds, and Grammar* (Chicago, 1996).

9 The analysis of the system of reference and co-reference encoded in specific texts is the central interest of the Theory of Mental Spaces. See Elena Semino, *Language and World Creation in Poems*

of discourse rather than individual sentences.[10] The idea of mental worlds has also been explored in Cognitive Science, where a distinction is drawn between the processing of pieces of information (propositions) and systems of meaning (models): P. N. Johnson-Laird's Theory of Mental Models[11] and Richard Gerrig's idea of being 'transported' to another world during the act of reading[12] both rely on this idea in their analysis of the psychological process of reading. Similarly, neuroscientists are finding that neutrally based 'scene construction' during the act of imagination involves the 'generation, maintenance and visualization of complex spatial contexts'.[13] For literary scholars, Text World Theory is especially useful as it tracks how mental representations are created and maintained over extensive pieces of fictional narrative discourse. Text World Theory also draws on the Possible Worlds Theory used by narratologists to account for the way readers compare the relative truth value of propositions, scenarios and events in narrative texts and those in the real world, but is distinguished by its language focus on discourse processing.[14] It also bears many similarities with Catherine Emmott's study of how people read narratives, in which she uses a cognitive discourse processing approach to analyse the structure and connectedness of longer texts, especially 'the demands which texts place on their readers in terms of inference-making and the mental storage of information'[15] (in this case through a close study of pronouns). Emmott's later work, which has grown out of this research into the 'mechanics of basic inference-making',[16] has been embraced within Cognitive Poetics, particularly her 'contextual frame theory',[17] and so now forms part of its interpretive range.

The point of contact among these approaches to the complex human cognitive reaction to language lies in the acknowledgement that fictional texts provide the opportunity for readers to imagine alternative realities that

and Other Texts (London, 1997); 'Possible Worlds and Mental Spaces in Ernest Hemingway's "A Very Short Story"', in *Cognitive Poetics in Practice*, ed. Joanna Gavins and Gerard Steen (London and New York, 2003), pp. 83–98; and 'Blending and Characters' Mental Functioning in Virginia Woolf's "Lappin and Lapinova"', *Language and Literature* 15 (2006), 55–72. See also E. Semino and J. Culpeper, eds, *Cognitive Stylistics: Language and Cognition in Text Analysis* (Amsterdam, 2002); and Todd Oakley and Anders Hougaard, eds, *Mental Spaces in Discourse and Interaction* (Amsterdam and Philadelphia, 2008).

10 Werth, *Text Worlds*, p. 77.

11 P. N. Johnson-Laird, *Mental Models* (Cambridge, 1983).

12 Richard J. Gerrig, *Experiencing Narrative Worlds: On the Psychological Activities of Reading* (New Haven and London, 1993), esp. pp. 2–17. Gerrig, who builds on the work of Johnson-Laird and others, examines the psychological structures underlying the phenomenon of experiencing transportation when reading particular narratives, and describes the 'situational models' or worlds that provide the organising principle for the complex reasoning that occurs during that process. One of the most interesting aspects of Gerrig's approach is his ability to account for 'participatory response' (p. 65), or emotional engagement with the narrative world.

13 D. Hassabis, D. Kumaran and E. A. Maguire, 'Using Imagination to Understand the Neural Basis of Episodic Memory', *The Journal of Neuroscience* 27.52 (2007), 14365–74, at p. 14365.

14 See Marie-Louise Ryan, *Possible Worlds, Artificial Intelligence, and Narrative Theory* (Bloomington, 1992), and Ruth Ronen, *Possible Worlds in Literary Theory* (Cambridge, 2004).

15 Catherine Emmott, *Narrative Comprehension: A Discourse Perspective* (Oxford, 1997), p. xi.

16 Ibid., p. 69.

17 Catherine Emmott, 'Reading for Pleasure: A Cognitive Poetic Analysis of 'Twists in the Tale' and Other Plot Reversal in Narrative', in *Cognitive Poetics in Practice*, ed. Gavins and Steen, pp. 145–59, at p. 146.

nevertheless bear some relationship to the real world, even if it is a contrastive one, and to create and update those models over the course of the text. The interaction of the two systems – the system of knowledge contained within in the narrative text and similarly systematic knowledge available from the real world – relies on the reader's understanding of what is considered to be possible within experienced reality. If events that are not considered to be possible in the real world occur, such as a person being in two places at the one time, then an impossible world is identified (with implications for genre, and therefore, readerly expectations, as in the case of science fiction). This philosophical theory, arising from theories of truth based in modal logic, thereby goes beyond standard narratology in providing a systematic consideration of the non-factual states of affairs, and a way of analysing the interplay of the normal, the possible and the impossible in narrative fiction.[18]

Text World Theory deploys this logic of Possible Worlds Theory to go beyond a consideration of the relative possibility of literary scenarios to an analysis of the conceptual processes that occur during meaning-making within all forms of communication. As a discourse framework, Text World Theory takes a broader approach to the examination of context than other theories, and makes a considerable contribution to our understanding of the experience of reading. This approach is language-based, but also takes account of universal cognitive functioning, and the contingent influence of cultural context, shared assumptions about human behaviour, and specific image schemas.

This theory is still being developed, especially in its application to literary narrative, yet is already highly detailed and technical in its specific applications. As an introduction to this new terrain, a simplified digest of some of its key points is provided immediately below, along with some suggestions for how the theory might be applied to Old English poetry.

Key concepts

There are several core concepts of Text World Theory.[19] Foremost among these is the distinction between two conceptual levels, the *discourse world* – 'the situational context surrounding the speech event itself',[20] and the *text world* – 'the situation depicted by the discourse'.[21] In the case of a written narrative text, the participants of the discourse world are the authors and readers, and in the text world those participants are (fictional) characters. For example, in the *Wulf and Eadwacer* citation above, the discourse world is the context in which the printed poem is read, and the text world is the speaker's context within that text, with its own subordinate or attached worlds created through memory and longing.

18 Luc Herman and Bart Vervaeck, *Handbook of Narrative Analysis* (Lincoln and London, 2005), p. 159.

19 The discussion following is only a brief list of the central ideas. For further details, see Werth, *Text Worlds*, and Gavins, *Text World Theory*, discussed above, nn. 3 and 5.

20 Werth, *Text Worlds*, p. 83.

21 Ibid., p. 87.

Text World Theory also relies on the idea of *common ground*, or 'the totality of information which the speaker(s) and hearer(s) have agreed to accept as relevant for their discourse'.[22] These mentally accumulated propositions of the discourse are held as assumptions by both parties (and guessed at in relation to each other), and their relevance to the discourse at hand is likewise implied. This silent and approximate co-operation between the writer and reader leaves not only a great deal of wiggle-room in terms of relevance, but also huge scope for various selection and deployment of ideas and, in turn, interpretation. In the *Wulf and Eadwacer* example above, that shifting set of imagined premises accounts for a huge portion of the interpretive variation relating to this text. This group of ideas evolves as the discourse progresses, changing as information is provided and new concepts and images introduced. This process of *incrementation* means that each new piece of information is managed to cohere with and to modify the common ground, as the text world is updated. Because most of this information is implied and called upon rather than explicitly stated, it is useful to imagine the text as a 'network of inference',[23] and for reading to be seen as the activation of that deductive process of sense-making, and also as a process that requires sharing a set of propositions relevant to a specific text. Indeed, the text drives the specific construction of context, which is negotiated by the reader and writer through the process of incrementation. The way textual clues act to trigger that process of *text-driven* coherence is a guiding principle of Text World Theory, which sees the accumulation of information as being determined by its immediate relevance to the textual context. Just how the text governs the selective activation of knowledge is a matter of keen interest in Text World Theory.

Information in text worlds falls into two main groups: *world-building* elements that define the background (static, descriptive, scene setting pieces of information); and *function-advancing* propositions that express actions and processes that move the discourse forward (events, or statements of motion). This framework relies on the technical linguistic category of *deictic indicators* within the world-building elements (deixis refers to the capacity of language to indicate position, especially in time and space, as in 'then' and 'there'). These referential indicators are 'part of the modality function' of language, which has the capacity to point and refer to situate information in terms of the current context, especially in terms of space, but also in terms of probability (truth value), and participant interaction.[24]

The textual dynamic is divided along a different axis in the distinction between *figure* (foregrounded information) and *ground* (background). In a text, significant pieces of information are clustered or pattered through the relief provided by unmarked background ideas. Foregrounded information is important for creating textual patterning, but background information is significant too, for the provision of privileged context, which itself can change through

22 Ibid., p. 119.
23 Ibid., p. 122.
24 Ibid., p. 157.

accommodation ('the presentation of new information in a background way'[25]). Sometimes a complete reassessment and reorientation is required by the reader, in view of incremental shifts in emphases and the emergence of new priorities for sense-making. Here, *world repair* is required, where the reader reinterprets and reconfigures assumptions in light of new information, and dismantles some of the incrementation that comes to seem to have occurred in error.[26]

Text worlds are *rich*, in that they are life-like, with characters that are created by the reader to be like real people. They represent human experience in complex and detailed ways, since they call upon cognitive models and are filled out with information from the reader's own memories, knowledge and conceptual schemas. They therefore have a large degree of 'internal structure' that requires active creation and negotiation.[27] This internal structure allows the text word to be further broken up into *sub-worlds* ('world switches' in Gavins's version of the framework),[28] which are different aspects or perspectives marked by a shift in time or space, triggered by a flashback, or a movement of the text into the mind of a particular character, and also to mark wishes and obligations ('boulomaic' and 'deontic' modal worlds in Gavins's terminology).[29] These switches involve cognitive repositioning, so are therefore noticeable and marked as potentially significant, because when a world switch occurs, a reader must reorient themselves both spatially and temporally in response to the new deictic markers. These sub-worlds are embedded in the main text world, and can be grouped into those that are able to be accessed and therefore verified by characters within the text world (*character-accessible sub-worlds*), and those verifiable directly by readers (*participant-accessible sub-worlds*). This distinction is important because it separates what can be verified as coming from the narrator and what has been filtered through a character, in much the same way as narrative focalisation is seen to work.

An important feature of this discourse framework in relation to fictional narrative is its recognition of a text's capacity to reorient a reader's perspective via a switch to a new time and place. When the textual here and now moves to a new position, the reader has to *project* their viewing position (*deictic centre*) into the new sub-world, in order to keep track of what is going on. This repositioning is part of a new conceptualisation (text-world or sub-world) that must be created in the mind. Gavins explains this process as the projection of a reader's 'notion of a zero reference point onto someone or something else in the text world' as a form of 'imagined transportation',[30] a similar concept to that developed by both Johnson-Laird and Gerrig mentioned above.

As in other cognitive paradigms, metaphor holds special significance. In this framework, *metaphors* are treated as a form of *double vision*. The major

25 Ibid., p. 280.
26 Gavins, *Text World Theory*, p. 141.
27 Werth, *Text Worlds*, p. 204.
28 See Gavins, *Text World Theory*, esp. p. 52, for details. Gavins prefers 'world-switch' to 'sub-world', as she regards the latter term as restrictive in its suggestion of subordinate connection with the main text world.
29 Ibid., pp. 94 and 98.
30 Ibid., pp. 40–1.

conceptual metaphors of a culture, or *megametaphors*, are significant, as they underpin not only propositions of the common ground, but also the systematic networking of ideas and assumptions that functions within textual patterning and conceptual connections. For example, the idea of knowledge as fruit is a megametaphor treated below, pp. 92–3, in the discussion of *Genesis B*, where metaphors of knowledge as vision are also discussed.

Text World Theory as an interpretive technique

Because it arises as a linguistic framework, and attempts to account for all human communication, Text World Theory usually focuses on modern English examples, and often treats texts other than literary ones. To date, little use of this theory has been made in relation to pre-modern literature, though there are some real benefits of this model of textual processing available to the Anglo-Saxonist. For instance, the probability that the present-day reader will share less common ground with the producer of the discourse world than the contemporary textual recipient – and also acknowledge and try to accommodate that deficit of cultural knowledge – deserves scrutiny, especially in terms of implications for meaning. Even more interesting is how this theory can account for the potential for emotional similarity: how we can be moved by texts written so long ago.

Peter Stockwell addresses precisely these issues in relation to a range of modern and early-modern literary texts. He extends the Text World Theory and combines it with other cognitive poetic techniques in his analysis of the role of emotion in the aesthetics of reading, or 'texture': 'the experienced quality of textuality'.[31] This more literary development of the theory is particularly useful in that it seeks to account for the mechanisms of literary reading: how certain ideas and knowledge from a reader's whole store of information are selectively recalled and deployed in response to textual cues in a particular act of reading; and how texts foreground and group certain ideas, inviting certain associations and cumulative responses in the mind of the reader. Specifically, Stockwell has developed a stylistic approach that accounts for the literary creation of ambiguity, tone and the affective reading experience. He looks at how the reading mind is primed below the level of consciousness to create certain connections in the text at hand and to derive a particular sort of reading experience from that text.[32]

In his detailed analysis of textuality and his account of various effects of literary reading, Stockwell makes more extensive use of Langacker's Cognitive Grammar than does either Werth or Gavins. In particular, Stockwell relies on the way that 'cognitive grammar explains how clauses are construed on the basis of a windowing of attention of different parts of the clause, a process called profiling'.[33] Stockwell uses Langacker's metaphor of 'force dynamics'

31 Peter Stockwell, *Texture: A Cognitive Aesthetics of Reading* (Edinburgh, 2009), p. 1.
32 Ibid., p. 11.
33 Ibid., p. 10.

within the clause, which entails energy transfer in an 'action-chain' – as in the case of one billiard-ball colliding with another, where the energy is transferred from the first to the second ball – to map the process of profiling. Certain ideas are made more prominent than others, and in turn drive the creation of further prominence: they create a 'construal effect' in the mind of the reader through systematic profiling. Stockwell analyses how construal varies, depending on linguistic form and textual organisation: how 'the propositional content of a piece of language is experienced in a variety of ways, depending on the manner in which it is presented and viewed'.[34] He approaches this question of the experience of meaningfulness by paying close attention to the systematic profiling and clustering of ideas in texts.

Stockwell's technical literary-linguistic approach contributes many valuable ideas. Some of the most interesting for the Anglo-Saxonist include the literary properties and linguistic creation of 'resonance' ('a sense of powerful literary impact and its ability to linger in the mind after the reading is completed'[35]); 'psychological projection', where readers model the minds of characters;[36] and the 'ethical positioning' of readers by texts.[37] More specifically, he deploys the psychology of attention to focus on the link between attention and meaningfulness, with 'attractors' that capture the attention,[38] either through novelty or stylistic foregrounding (e.g. through motion or contrasts in brightness and noisiness).[39] He explores the way that texts use attractors and the reverse, 'attentional neglect' (backgrounding) to direct the reader's attention to particular patterning elements through 'attentional drag and zoom'.[40] An interesting result is the exploitation within textual organisation of the phenomenon of 'attentional blink', or a blind-spot where content is occluded into the background after an instance of invited scrutiny by the reader.[41] The distinction made between a figure (an object of attention) and ground (background) invites the extension of the metaphor to the significance of the outline or edge of the figure, especially one that passes through the horizon of attention during reading: the impact of moving edges. Stockwell takes particular notice of the way the figures move into and out of our field of vision as we read, so as attractors are dynamic; in turn they pass on that sense of motion to other attractors and thereby drive textual focus.[42]

Stockwell's aim is to develop a literary-critical approach from the linguistic cognitive poetic perspective, to synthesise methods from the fields of Stylistics and Literary Criticism in order to create fine-grained, formal analytical tools to account for how textual form and organisation feed into interpretation, aesthetics and readerly affect. As he does so, he demonstrates the degree

34 Ibid., p. 171.
35 Ibid., p. 14.
36 Ibid.
37 Ibid., p. 7.
38 Ibid., p. 20.
39 Ibid., p. 25.
40 Ibid., p. 43.
41 Ibid., p. 67.
42 Ibid., p. 40.

of overlap between Cognitive Poetics and Cognitive Cultural Studies. His discussion of Theory of Mind, or mind-modelling, for example, shares ground with, and cites the work of Lisa Zunshine,[43] discussed above. This overlap suggests the flexible and fruitful lines of inquiry that cognitive approaches can offer, especially in view of the potential to adapt interpretive tools selectively rather than to adopt a new, fully fledged interpretive paradigm.

The point at which these lines of inquiry becomes particularly interesting for the Anglo-Saxonist is in the consideration of how literary texts can exploit an evolved, embodied human universal: the connection between physical experience and conceptualisation. This connection gives rise to image-schemas that were meaningful to the Anglo-Saxon poet and audience, as well as to modern readers. For instance, the schema of containment, with connotations for imprisonment, privacy and safety, is evident in *Wulf and Eadwacer*; and *Genesis B* deploys a recognisable image-schema of knowledge as vision to drive the cumulative prominence of particular ideas. Furthermore, texts can invite intense, involved readings from a culturally remote, modern audience because they evoke and thereby produce emotional experiences and genuine sensations, notwithstanding their acknowledged fictionality. Stockwell is broadly in agreement with Gerrig's argument that recognised fictionality does not affect actual cognitive processing: 'the experience of narratives is largely unaffected by their announced correspondence with reality'.[44] Therefore, ethical positioning, readerly implication, or emotional investment, arises from rich, textured features within discourse – varied collections of attractors and 'elegant variation of the patterning'[45] – which are recognised features of Old English poetry.

The Text World framework and its deployment in Stockwell's stylistic analysis of the 'cognitive basis of both feeling and meaning'[46] are applied below in relation to three Old English texts differing in type – *Beowulf*, *Genesis B* and *Wulf and Eadwacer* – in order to demonstrate its potential for literary interpretation, and its capacity to play a role in the development of less specifically linguistic cognitive approaches to literature with a focus on Old English texts.

Text worlds in *Beowulf*

The hero's first speech to Hrothgar on his arrival at Heorot near the beginning of *Beowulf* (407–55) offers an opportunity to demonstrate some of these ideas in practice in a familiar text. In particular, it provides an interesting example of multiple, rich text worlds, the management of several complex timelines and textual layers via deictic indicators, and the foregrounding and clustering of particular attractors, to create a construal effect, or focus on prominent ideas. In this brief excerpt Beowulf explains how he came to hear

43 *Texture*, pp. 139–44.
44 Gerrig 102
45 Stockwell, *Texture*, p. 70.
46 Ibid., p. 4.

of Grendel's menace, and how he himself is equipped to deal with the situation. The latter part of the speech, which describes what will happen should Beowulf fail in the proposed battle with the monster, serves to reinforce his asserted intention to succeed rather than otherwise, a literary effect that is open to close analysis. In negotiating this piece of text, the reader is inclined to implicate him or herself in the text world and to create a positive mental representation of Beowulf as hero, an act of ethical positioning triggered by the mental processing of the text.

The following discussion is designed to suggest how the Text World approach can explain this process, and thereby to show its capacity to extend areas of discussion that are already occurring in *Beowulf* scholarship, especially on the topics of speech and conceptualisation. For instance, several scholars have analysed the use of direct discourse in the poem,[47] the textual interest in perception,[48] and creation of suspense via modulated point of view.[49] The disjunctive timeline, with embedded memories, narratives and ominous flashforwards (for instance, the reference to the burning of Heorot when its creation is being described) in this text have likewise been well canvassed.[50] The associative lexical style of this poem has been subject to close and convincing scrutiny.[51] Its speeches, which comprise about forty percent of the poem, have been subject to many types of critical investigation,[52] including in the field of Discourse Analysis.[53] More specifically, Beowulf's speech at lines 407–55 has itself been the subject of close study, in terms of its

47 Robert E. Bjork, 'Speech as Gift in *Beowulf*', *Speculum* 69 (1994), 993–1022, which surveys the earlier scholarship and includes a useful set of appendices of speeches and their stylistic features.

48 Peter Clemoes, 'Action in *Beowulf* and Our Perception of It', in *Old English Poetry: Essays on Style*, ed. Daniel G. Calder (Berkeley, 1979), pp. 147–68; Margaret A. L. Locherbie-Cameron, '"Anglo-Saxon Attitudes": The Visual Nature of Some Poetic Narrative Structures', *Parergon* 10.2 (1992), 71–82; Ward Parks, 'Prey Tell: How Heroes Perceive Monsters in *Beowulf*', *Journal of English and Germanic Philology* 92 (1993), 1–16; Michael Lapidge, '*Beowulf* and Perception', *Publications of the British Academy* 111 (2001), 61–97 (Sir Israel Gollancz Memorial Lecture).

49 Alain Renoir, 'Point of View and Design for Terror in *Beowulf*', *Neuphilologische Mitteilungen* 63 (1962), 154–67; Gail Ivy Berlin, 'Grendel's Advance on Heorot: The Functions of Anticipation', *Proceedings of the PMR Conference* 11 (1986), 19–26; Michael Lapidge, '*Beowulf* and the Psychology of Terror', in *Heroic Poetry in the Anglo-Saxon Period: Studies in Honor of Jess B. Bessinger*, ed. Helen Damico and John Leyerle, Studies in Medieval Culture 32 (Kalamazoo, MI, 1993), pp. 373–402. David Sandner, 'Tracking Grendel: The Uncanny in *Beowulf*', *Extrapolation* 40 (1999), 162–76; and William Perry Marvin, 'Heorot, Grendel, and the Ethos of the Kill', *In Geardagum* 24 (2003), 1–39.

50 For example, A. Bonjour, *The Digressions in 'Beowulf'* (Oxford, 1950); and Norman E. Eliason, 'The Burning of Heorot', *Speculum* 55 (1980), 75–83.The accumulated literature on Beowulf is particularly large. Critical material on this text, as on other Anglo-Saxon topics, is available on the *Old English Newsletter* database: http://www.oenewsletter.org/OENDB/index.php.

51 Fred C. Robinson, *Beowulf and the Appositive Style* (Knoxville, 1985).

52 For a pithy summary of scholarship on direct speech in *Beowulf*, see R. D. Fulk, Robert E. Bjork and John D. Niles, eds, *Klaeber's Beowulf and the Fight at Finnsburg*, 4th edn (Toronto, 2008), pp. lxxxvi–lxxxviii.

53 Dolores Fernández Martínez, 'The Study of Old English Literary Texts from the Perspective of Discourse Analysis: An Overview', *The Grove: Working Papers on English Studies* 8 (2001), 69–79.

syntactic structure[54] and rhetorical devices.[55] Since this speech is one of the places in the poem in which Beowulf's youthful adventures are recounted, it also exemplifies the global narrative reliance on so-called 'digressions' from the central narrative in the poem, noted and discussed by many scholars.[56] Indeed, it is difficult to improve on John Niles's summary comment on the poem: 'an abstract, non-representational work of art that splits and recombines the elements of everyday perception to show the spiritual or ethical relations that lie behind things'.[57] These lines of inquiry can be developed further through a consideration of the interpretive processes triggered by the text: how it primes the recipient to conceptualise multiple textual frames and to configure meaning from stated and implied information, and from contributed perceptual and experiential knowledge.

The text world of *Beowulf* is rich and structured, with multiple embedded sub-worlds or world switches. The digressions and sub-narratives of this long poem can be seen in this light as a collection of several different discourse worlds depicted within the overarching text world. The slippage among these worlds results in uncertainty in the reader's mind about precise world boundaries. But even more pervasive than the narrative digressions – and forming a platform for them – is the practice of routine multiple sub-world creation within the fabric of the text, a process that drives conceptual association and ultimately shapes textual interpretation. Besides these shifts, there is a more fundamental change of scene in the middle of the long speech, which is divided below into two parts to facilitate discussion.

Beowulf's first speech to Hrothgar: part 1 (407–426a)

'Wæs þu, Hroðgar, hal!   Ic eom Higelaces
mæg ond magoðegn;   hæbbe ic mærða fela
ongunnen on geogoþe.   Me wearð Grendles þing
on minre eþeltyrf   undyrne cuð;
secgað sæliðend   þæt þæs sele stande,
reced selesta   rinca gehwylcum
idel ond unnyt,   siððan æfenleoht
under heofenes haðor   beholen weorþeð.
Þa me þæt gelærdon   leode mine
þa selestan,   snotere ceorlas
þeoden Hroðgar,   þæt ic þe sohte,
forþan hie magenes cræft   min[n]e cuþon;

54 Thomas J. Jambeck, 'The Syntax of Petition in *Beowulf* and *Sir Gawain and The Green Knight*', *Style* 7.1 (1973), 21–9.

55 E. Irving, *A Reading of Beowulf* (New Haven, 1968), pp. 62–8. John Kenny Crane, 'To Thwack or Be Thwacked: An Evaluation of Available Translations and Editions of *Beowulf*', *College English* 32 (1970), 321–40, uses part of this speech (lines 407–32), as his basis on which to compare the twelve translations he reviews.

56 John D. Niles, *Beowulf: The Poem and its Tradition* (Cambridge, MA, 1983), pp. 163–76; and Robert E. Bjork, 'Digressions and Episodes', in *A Beowulf Handbook*, ed. R. Bjork and J. Niles (Lincoln, NE, 1997), pp. 193–212.

57 Niles, *Beowulf: The Poem and its Tradition*, p. 176.

selfe ofersawon, ða ic of searwum cwom
fah from feondum, þær ic fife geband,
yðde eotena cyn, ond on yðum slog
niceras nihtes, nearoþearfe dreah,
wræc Wedera nið – wean ahsodon –
forgrand gramum; ond nu wið Grendel sceal,
wið þam aglæcan ana gehegan
ðing wið þyrse.'

'Hail to you, Hrothgar! I am Hygelac's kinsman and young retainer. In my youth I have undertaken many glorious deeds. The matter of Grendel became known to me on my native soil; seafarers say that this hall, best of buildings, stands idle and useless to all warriors once the evening light becomes hidden under the confinement of heaven. Then my people, the best wise men, advised me, lord Hrothgar, that I should seek you, because they knew the power of my strength. They themselves had observed it when I, stained with the blood of enemies, returned from the contest, bound five, killed a race of monsters and in the waves slew water-monsters by night; I suffered harsh times, avenged the wrong of the Weders, completely ground down those creatures – they sought trouble. And now I alone am determined to settle this matter with the monster Grendel.'

In making sense of this excerpt, the reader is called upon to negotiate a good deal of expressed and implied common ground, multiple world shifts and the chain of attractors clustered around action, success and resolution communicated through Beowulf's associative logic. In this except, information is presented as a part of the common ground of Beowulf's home context, rather than as a matter of his opinion: the use of firm indicative statements rather than modalised ones creates a set of confident assertions of truth rather than of opinion ('I am', 'they knew', 'I killed'). It treats the past as preparation for the present challenge ('now'), a context in which the existence of monsters, and in turn heroes to defeat them, is part of the established state of affairs. This expressed and implied common ground is in keeping with the way it is invoked right from the outset of the poem, with the addressee potentially included in the plural pronoun used by the 'authenticating voice' of the narrator:[58] 'We have heard of the glory of the Spear Danes' (1). Even if this reference and those within Beowulf's speech are allusive for the modern reader without some knowledge of Anglo-Saxon history, culture and literary style, they immediately create a referential frame that forces the reader towards interpretive multi-tasking in order to keep up and create coherence.

The most noticeable component of this textual complexity is temporal: the narrative time sequence is disjointed, a textual feature that can be mapped. As Fig. 1 shows, the multiple embedded world switches require the reader to keep track of several layers of sub-worlds and directed clusters of ideas. Most noticeably, once Beowulf begins to speak, all the sub-worlds are filtered through his perception, so are character-accessible only. The reader, like his

[58] Stanley B. Greenfield, 'The Authenticating Voice in *Beowulf*', *Anglo-Saxon England* 5 (1976), 51–62.

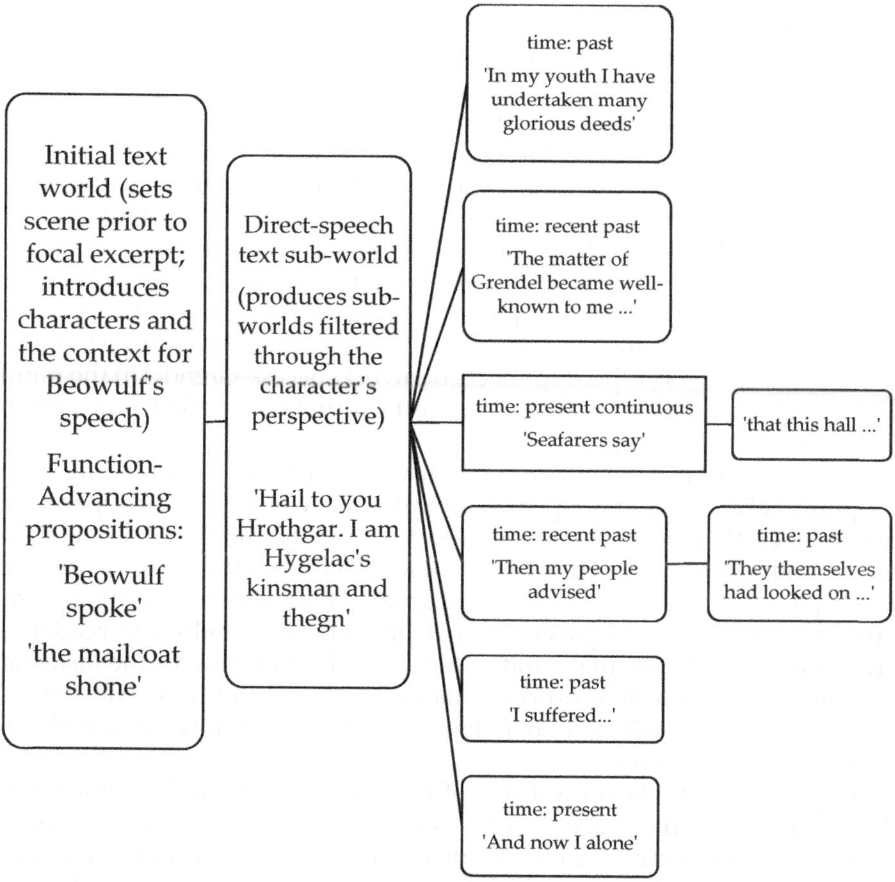

FIG. 1. Text worlds and sub-worlds in *Beowulf*, lines 407–426a

immediate fictional audience, including Hrothgar, are dependent on him for the information and accuracy of presentation, and on his reliability as a source of information. Moreover, the reader is called up to project their viewing position onto the sub-worlds that spring from Beowulf's memory. His audience is similarly reliant on him for the sequencing and organisation of information. For example, he presents his credentials and details of how he came to hear of the problem before promising his services to address it. Significantly, he says that he is acting as a result of the speech of others: seafarers told him of Grendel, and he, Beowulf, was advised to seek out Hrothgar to help. All the sub-worlds are centred on Beowulf, the recipient of information from others, and/or Grendel, the subject of that information, so they develop a connection between the two. The result is a degree of awareness of the text's 'attentional drag and zoom',[59] whereby the reader's attention is heavily managed by the

59 Stockwell, *Texture*, p. 43.

shifting deictic references and changing sub-worlds of the text. The text establishes Beowulf as a victor in the hero-versus-monster schema it calls up, and a righteous defender of the persecuted, in a construal effect that accumulates focus on his success in the forthcoming battle.

The built-up textual attractor, Beowulf as hero, is conceptualised by the reader before the pronouncement at the conclusion of this section, 'and now I alone am determined to settle this matter with the monster Grendel'. The 'and' encodes a causal connection between the relevant credentials just enunciated and the assertion of success against the current monster. The use of the word 'alone' picks up on the successful exploits of the solitary hero just related and projects Beowulf more emphatically as an outstanding individual. The action chain connects Beowulf-success-Grendel in the mind of the reader, a projected future state of affairs that is ostensibly qualified in the second part of the speech.

This passage thereby persuasively invites a 'prototypical reading' that cooperatively accepts the speaker's expressed views.[60] This cooperation results not only from the reader's collection and synthesis of accumulated information, but as a result of being invited to project their minds into the hero's past and present, poised to imagine certain outcomes in his future. The result is a form of psychological projection whereby the reader is invited to model Beowulf's mind in terms of his readiness for the task and his commitedness to it, a process that is developed in the second part of the speech as the chain of associations takes the reader towards a belief in Beowulf's future success.

In systematically keeping track of the propositions as they emerge in this piece of text, the reader imagines and updates the text world, makes inferences, and synthesises those ideas with customised selections from information stored in memory. The way this process is driven by the text can be mapped (see Fig. 2).

Beowulf's first speech to Hrothgar: part 2 (426b–455)

'Ic þe nuða,
brego Beorht-Dena, biddan wille,
eodor Scyldinga, anre bene,
þæt ðu me ne forwyrne, wigendra hleo,
freowine folca, nu ic þus feorran com,
þæt ic mote ana minra eorla gedryht,
ond þes hearda heap, Heorot fælsian.
Hæbbe ic eac geahsod þæt se aglæca
for his wonhydum wæpna ne recceð;
ic þæt þonne forhicge, swa me Higelac sie,
min mondrihten modes bliðe,
þæt ic sweord bere oþðe sidne scyld,
geolorand to guþe, ac ic mid grape sceal

60 Ibid., p. 162.

fon wið feonde   ond ymb feorh sacan,
lað wið laþum;   ðær gelyfan sceal
dryhtnes dome   se þe hine deað nimeð.
Wen' ic þæt he wille,   gif he wealdan mot,
in þæm guðsele   Geatena leode
etan unforhte,   swa he oft dyde,
mægenhreð manna.   Na þu minne þearft
hafalan hydan,   ac he me habban wile
d[r]eore fahne,   gif mec deað nimeð
byreð blodig wæl,   byrgean þenceð,
eteð angenga   unmurnlice,
mearcað morhopu –   no ðu ymb mines ne þearft
lices feorme   leng sorgian.
Onsend Higelace,   gif mec hild nime,
beaduscruda betst,   þæt mine breost wereð,
hrægla selest;   þæt is Hrædlan laf,
Welandes geweorc.   Gæð a wyrd swa hio scel.'

'Now, leader of the Bright-Danes, lord of the Scyldings, I want to ask for one request from you: that you, protector of warriors, dear friend of people, do not refuse me, now that I have come this far, that I alone, along with my company of warriors, this troop of hardy men, may cleanse Heorot. I have also learned that out of recklessness this monster carries no weapons. Therefore, so that my lord Hygelac may be happy in his mind, I disdain to carry a sword or a broad shield, a yellow disc, into battle, but I am determined to come to grips with the enemy and fight to the death, foe against foe. He whom death takes must trust in the judgement of the Lord. I imagine that, if he succeeds, he will fearlessly consume the people of the Geats, the mighty host, in the war-hall, as he has often done. There will be no need for you to cover my head, but he will have me stained with blood, if death takes me the lone stalker will carry away my bloody corpse, will intend to devour it, to eat it mercilessly, will stain his retreat in the moors. There will no longer be a need for you to worry about taking care of my body. Send Hygelac, if war should take me, the best of battle-gear that protects my breast, the most excellent of garments, that is a legacy of Hrethel, the work of Weland. Ever goes fate as she must!'

This part of the speech foregrounds the addressee through an appositive cluster for Hrothgar at the outset ('Beorht-Dena, . . . eodor Scyldinga, . . . wigendra hleo', 427–9), creating attractor status for the king as the recipient of Beowulf's rhetorical pitch, and cuing a referential frame that features respectful speech-making and the display of credentials. Imaginary propositions play a far greater role in this part of the speech than the other. After the address at the opening, this second part of the speech, mapped in Fig. 2, contains far more modalised sub-worlds than the first, creating a tactful tentativeness.[61] But, building upon the argument that the syntactic manner of Beowulf's speech serves the social and rhetorical functions of neutralising any opposition to his plan in advance, a Text World approach can explain the switches commencing with 'I wish', 'I may', I imagine' and the conditional

61 As noted by noted by Jambeck, 'The Syntax of Petition'.

'if' clauses, as the projection of future scenarios embedded within Beowulf's speech. Through the syntax of the sentence, the reader has to create mental representations of these modal sub-worlds in order to make sense of the text, and thereby recalibrates the conceptual relationships among propositions, setting up a chain of inference-making that leads to a certain type of textual coherence, resonance and ethical positioning.

The collections of sub-worlds generated by this passage work with the firmer indicative statements of the first section to attract solidity. The final, gnomic utterance, 'Ever goes Fate . . .' lifts the level of abstraction to the extreme, though attains relevance and traction through its place in the sequence starting with biography, moving to intention, then to likelihood and finishing with this assertion from an apparently reliable speaker. In Text World terms, the reader is able to create a text world combining Beowulf and the concept of fate at the conclusion of this speech, a pairing that is reiterated elsewhere in the poem, as a direct result of being led from the past exploits of Beowulf to his new intentions through the energy transfer facilitated by the shifting text worlds. This experiential investment is reinforced as the reader is encouraged to synthesise the attributes Beowulf collects from the world-switches creating momentum for a conceptual trajectory of Beowulf + success.[62]

This process becomes particularly interesting at the end of the speech, where Beowulf outlines a scenario of failure. This part of the speech is introduced by the world switch into the gnomic utterance 'He whom death takes must trust in the judgement of the Lord.' Then, the reader is invited into Beowulf's imagination, where Grendel's success is entertained provisionally, setting up a negative possible world, which is actually a continuation of the present state of affairs. Current reality thereby becomes an ongoing 'horror scenario' if the hero fails. The tag line 'as he has often done' confounds the expression of this image as the product of the hero's imagination, but rather presents it as a continuation of the status quo: a split deictic reference to possible future and current situations. He even projects Grendel's intentions ('will intend to devour it, to eat it mercilessly') via a further sub-world. The inclusion of 'no longer' in 'There will no longer be a need for you to worry about taking care of my body' implies the current obligation of host to guest in a new sub-world of social custom, within a sub-world of future possibility, a juxtaposition with the proximate image of Grendel eating that same body that creates a 'grim joke'.[63] This humorous contrast created through opposed sub-worlds of host-courtesy and monster-carnage urges the creation of an inference in the mind of the reader that defeat is mentioned lightheartedly or ironically as an impossibility. In Text World terms, the discursive force of the sentence acts against its propositional content, as the 'defeat-by-Grendel' sub-world is presented as shallow, or not fully formed, by virtue of the humour or at least the grim irony surrounding it.

62 Stockwell, *Texture*, p. 149.

63 Fulk, Bjork and Niles, *Klaeber's Beowulf*, 'Introduction', p. cx.

The final request, to send the war-gear back to Hygelac, along with the world switch into the back-history of the garment and the stuff of legend, creates a rich sub-world where the reader is invited to consider the earlier successful battles in which the garment was used, and the grief and loss Beowulf's death would cause at home, and to combine this idea with the image of the undisappointed Hygelac, who hears that Beowulf has not infringed the noble rules of combat with an unarmed assailant. The reader incrementally combines this information to the final line of the speech ('Ever goes fate as she must!'). This line acts in concert with this conceptual combination of Beowulf and honour, enhanced by his association with legendary material, so that it is not only a gnomic statement of one of the poem's themes, but also allows the attractor, fate, to be brought back to Beowulf, after it was set off in the reference to Judgement several lines previously.

If we consider these lines in terms of the 'force dynamics' outlined by Langacker and used by Stockwell, we can trace the arc of the metaphoric billiard ball of the attractor 'fate' to Grendel, back to Hygelac, and then to Beowulf himself, through 'my body', 'my breast', protected by legendary work by 'Hrethel'. Besides this emphasis on protection of the body, Beowulf collects several positive attractors, including individual responsibility for success, and a role in fated events. The reader is transported not only into the character's past and imagined future, but also into the mind of his adversary, his king at home, and into a neutral position of gnomic statement. The reader is also reminded of the mind/body distinction and the material present through the lexical emphasis on this pair of concepts within the multiple world creation of the extract (e.g. 'mod' 436b, 'hafela' 446a, 'lice' 451a and 'breost' 553b). The emphasis on Beowulf's physical body and the mental representation of it as a meal for a monster projects the reader's viewing position into the mind of a ravenous, ruthless monster, an uncomfortable repositioning that the rest of the speech undermines as a possible future state of affairs.

The gruesome image of Grendel devouring Beowulf that the hero's speech forces his audience to create mentally encourages an empathetic response in the reader, reinforced by the images of proud but grieving lord Hygelac back home. Similarly, the accumulation of images of success, created alongside a particularly unappealing image of defeat makes the success seem more rather than less likely. Through this series of projections by the speaker, attention is zoomed in on failure as a flimsy sub-world and its contrast to the multiple sub-worlds of success already created. The reader must conceptualise vague yet characteristic success and detailed singular defeat in order to make sense of this passage, and is thereby to reject the latter world, presented in its visual and experiential fullness, in favour of an asserted, but undetailed alternative. Grendel is thereby presented as a real threat, though one that is likely to be defeated, through the deictic shifts and presentation of the back-story create an expectation of success by Beowulf against that serious threat. This expectation is further reinforced by the way in which Beowulf is placed at the deictic centre of the world-switches and as the subject of fated events.

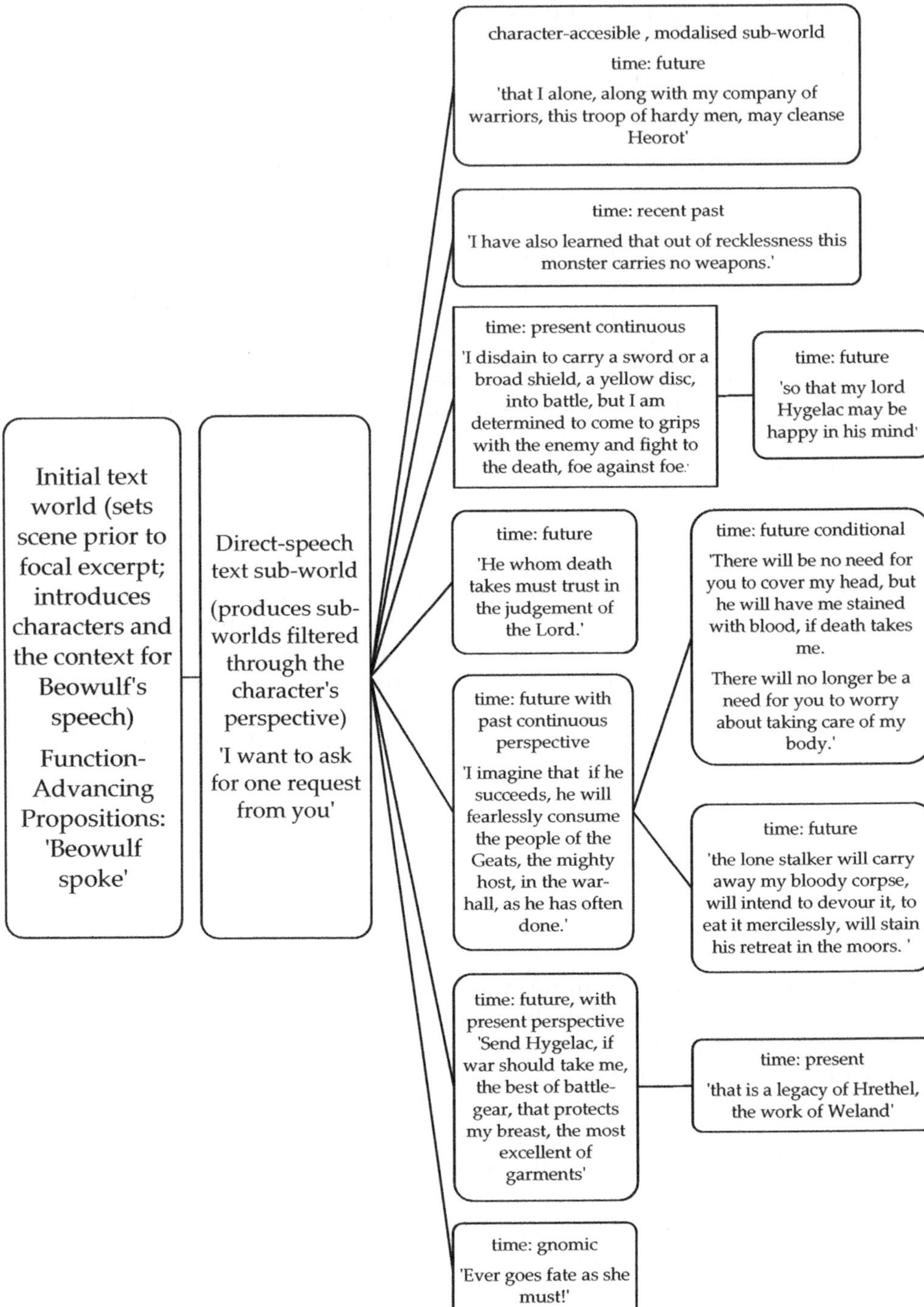

FIG. 2. Text worlds and sub-worlds in *Beowulf*, lines 426b–455

These ideas can combine well with existing scholarship on the speech, most notably the concept that in each of his three formal vows to fight against a monster, Beowulf uses speech as a form of 'social action' that allows him to take on the hero's role in this context.[64] Beowulf's first speech to Hrothgar achieves a complex effect that combines this social function with its ostensible persuasive function, to prime the reader to expect the hero's full control over the situation and his success, an effect achieved through the creation of multiple text worlds and time shifts that situate the hero at the centre of the past, present and future, indeed as the subject of fated events in a cosmic continuum. The physicality of his presence is embedded in the text, in contrast to the contents of his mind that are presented within the text worlds of his speech. As the first formal verbal encounter between the hero and the king for whom he will perform an act of bravery and liberation, this piece of communication creates in the minds of its audience both perceptions and experiences that place the hero alone in the imagined text world of success. The prominence of this idea within the speech is achieved through embedded and speculative ideas rather than through the assertions within the discourse, an effect that can be examined in a distinctly different sort of speech within the extant Old English corpus, in *Genesis B*.

Nested text worlds in Genesis B

The temptation of Eve in *Genesis B* (547–622) has received more than its share of critical attention, and has attracted especially polarised interpretive conclusions regarding the culpability of Eve and the rhetorical force of this scene. This divided critical reception of this text provides an ideal demonstration of the way multiple sub-worlds are embedded in both text and discourse worlds, influencing characterisation and interpretation in multiple ways. Text World Theory allows a way of systematically analysing the embedded wishes, unreal conditions and hypotheticals in the devil's temptation of Eve in *Genesis B* – the sub-worlds that the reader is called upon to construct as they negotiate the complex nesting of ideas in this passage – and in turn to account for the variety of critical readings the text has attracted. The following discussion will apply the theory of sub-worlds to this text in order to analyse the reliance of critical assumptions and inclinations on frames of mind and imputed motivations that are subjectively developed during the reading process.

In Text World Theory, readers are understood to develop, sustain and modify mental representations of the set of propositions made in narrative texts. Within this process, temporal and spatial shifts of focus within the narrative – sub-worlds – are negotiated by the reader. Sub-worlds are governed by different laws of logic and have different deictic parameters from those of their containing text world,[65] so can change the relationship among

64 Ibid., p. lxxxviii.
65 Werth, *Text Worlds*, p. 355.

the hierarchy of propositions. It is important to note the distinction between those sub-worlds that can be accessed by the reader (e.g. through a direct speech situation), and those that are directly available to a character, not the reader, yet are reported within the fiction (e.g. reported speech, memory).[66]

Scholars have long recognised a high degree of psychological realism in *Genesis B*, as initially discussed by Rosemary Woolf and Anne L. Klinck,[67] and developed by Eric Jager, who sees the Fall as the corrupting power of rhetoric on a susceptible mind.[68] But this apparent interest in the cognitive life is the source of the huge divergence in critical account of the characterisation and rhetorical force of this text. Most strikingly, the critical reception of *Genesis B* is characterised by a desire to deduce or imagine Eve's motivations and make a cognitive assessment of the causes and consequences of her actions. Critics have largely fallen into a pro- and anti-Eve division.[69] Those in the former camp, including Renoir and Burchmore, see Eve as a victim of circumstance and not culpable for the Fall,[70] and Davis, who sees her as suffering from the corruption of her senses.[71] On the other hand, those who see Eve as culpable in various ways include Ehrhert ('Eve should have noticed');[72] Woolf (failure of Eve's moral sense);[73] Vickrey (Eve 'a dupe')[74] Doane (who sees Eve as sinning through 'self-willed delusion');[75] Finnegan (Eve would

66 In *Text Worlds*, Werth divides these two types of sub-worlds – *participant-accessible* and *character-accessible* respectively – into several further categories that provide a level of technical distinction that is beyond the present demonstration (esp. p. 216).

67 Rosemary Woolf, 'The Fall of Man in *Genesis B* and the *Mystère d'Adam*', in *Studies in Old English in Honor of Arthur G. Brodeur*, ed. S. B. Greenfield (New York, 1963), pp. 187–99, at p. 198; and Anne L. Klinck, 'Female Characterisation in Old English Poetry and the Growth of Psychological Realism: *Genesis B* and *Christ I*', *Neophilologus* 63 (1979), 597–610, at p. 599. Psychology had also been of interest in Alain Renoir, 'The Self-Deception of Temptation: Boethian Psychology in *Genesis B*', in *Old English Poetry: Fifteen Essays*, ed. Robert P. Creed (Providence, 1967), pp. 47–67.

68 Eric Jager, 'The Tempter as Rhetoric Teacher: The Fall of Language in the Old English *Genesis B*', *Neophilologus* 72 (1988), 434–48; and 'The Word in the "breost": Interiority and the Fall in *Genesis B*', *Neophilologus* 75 (1991), 279–90. Jager's argument of the corrupting power of discourse was developed further in *The Tempter's Voice: Language and the Fall in Medieval Literature* (Ithaca and London, 1993), Chapter 4, 'The Old English Epic of the Fall', pp. 145–89.

69 The critical discussions up until 1990 are well rehearsed by Gillian R. Overing, 'On Reading Eve: *Genesis B* and the Readers' Desire', in *Speaking Two Languages: Traditional Disciplines and Contemporary Theory in Medieval Studies*, ed. Allen J. Frantzen (Albany, 1991), pp. 35–63 and 232–9. For later developments, see Susannah B. Mintz, 'Words Devilish and Divine: Eve as Speaker in *Genesis B*', *Neophilologus* 81 (1997), 609–23.

70 Alain Renoir, 'Eve's I. Q. Rating: Two Sexist Views of *Genesis B*', in *New Readings on Women in Old English Literature*, ed. Helen Damico and Alexandra Hennessey Olsen (Bloomington and Indianapolis, 1990), pp. 262–72; and Susan Birchmore, 'Traditional Exegesis and the Question of Guilt in the Old English 'Genesis B''', *Traditio* 41 (1985), 117–44.

71 Glenn M. Davis, 'Changing Senses in *Genesis B*', *Philological Quarterly* 80 (2001), 113–31.

72 Margaret J. Ehrhart, 'Tempter as Teacher: Some Observations on The Vocabulary of the Old English *Genesis B*', *Neophilologus* 59 (1975), 435–46.

73 Woolf, 'The Fall of Man in *Genesis B*'.

74 John Vickrey, 'The Vision of Eve in *Genesis B*', *Speculum* 44 (1969), 86–102; '*Genesis B* and the Anomalous Gnome', *Mediaevalia* 14 (1991 for 1988), 51–62; 'Adam, Eve, and the *tacen* in *Genesis B*', *Philological Quarterly* 72 (1993), 1–14; and '*Genesis* 549–51 and 623–5: Narrative Frame and Devilish Cunning', *Philological Quarterly* 76 (1997), 347–68.

75 A. N. Doane, ed., *The Saxon Genesis: An Edition of the West Saxon 'Genesis B' and the Old Saxon Vatican 'Genesis'* (Madison, WI, and London, 1991), p. 150.

have seen the tempter accurately);[76] and Langeslag (who argues that Eve sins out of loyalty to Adam).[77]

I would like to argue that the interest in this psychological aspect of the text, and the divergent interpretations of it, arise from the creation of distinct character-accessible sub-worlds, and in turn produce distinct schemas for the whole text world, which then colour the entire interpretive process. In order to analyse the tempter's speech closely and explore this argument more fully, the speech is presented below in its entirety, broken up into four main sections.

When he begins to seduce Eve into eating the fruit of the tree of knowledge, the tempter assumes full knowledge of the mind of God (551b–559a):

The tempter's speech to Eve: part 1

'Ic wat, inc waldend god
abolgen wyrð, swa ic him þisne bodscipe.
selfe secge, þonne ic of þys siðe cume
ofer langne weg, þæt git ne læstan wel
hwilc ærende swa he easten hider
on þysne sið sendeð. Nu sceal he sylf faran
to incre andsware; ne mæg his ærende
his boda beodan; þy ic wat þæt he inc abolgen wyrð,
mihtig on mode.'

'I know that the ruling god will be angered by you two, as I myself relate this message, when I return from this journey over the long path, that you two did not accomplish well whatever message he sends east hither on this occasion. Now he himself must travel for your answer; his envoy is not allowed to announce his message. Thus I know that he will become angry with you two, the mighty one in his mind.'

After presenting this potential (yet counter-factual) scenario, the tempter then goes on to instruct Eve, providing an irresistible future scenario conditional upon her eating the fruit. His method and the content of his message are both psychological: he plays on her mind by promising mind-powers in the form of divine vision. As his chief persuasive tactic, he encourages Eve to create a mental representation of enhanced visionary power (559b–563):

The tempter's speech to Eve: part 2

'Gif þu þeah minum wilt,
wif willende, wordum hyran,
þu meaht his þonne rume ræd geþencan.
Gehyge on þinum breostum þæt þu inc bam twam meaht
wite bewarigan, swa ic þe wisie.'

[76] Robert Emmett Finnegan, 'Eve and "Vincible Ignorance" in *Genesis B*', *Texas Studies in Literature and Language* 18 (1976), 329–39.

[77] P. S. Langeslag, 'Doctrine and Paradigm: Two Functions of the Innovations in *Genesis B*', *Studia Neophilologica* 79 (2007), 113–18. A similar line of argument is presented by Peter J. Lucas, 'Loyalty and Obedience in the Old English *Genesis* and the Interpolation of *Genesis B* into *Genesis A*', *Neophilologus* 76 (1992), 121–35.

'If you, wiling woman, will however obey my words, you might therefore consider his counsel extensively. Consider in your heart that you might avoid punishment for both of you as I instruct you.'

Having outlined the perils of ignoring his message through a conditional modal sub-world, and provided the rationale for regarding his instructions as divine and worth following, the tempter then stipulates the required course of action and its outcomes (564–80):

The tempter's speech to Eve: part 3

'Æt þisses ofetes! Þonne wurðað þin eagan swa leoht
þæt þu meaht swa wide ofer woruld ealle
geseon siððan, and selfes stol
herran þines, and habban his hyldo forð.
Meaht þu Adame eft gestyran,
gif þu his willan hæfst and he þinum wordum getrywð.
Gif þu him to soðe sægst hwylce þu selfa hæfst
bisne on breostum, þæs þu gebod godes
lare læstes, he þone laðan strið,
yfel andwyrde an forlæteð
on breostcofan, swa wit him bu tu
an sped sprecað. Span þu hine georne
þæt he þine lare læste, þy læs gyt lað gode,
incrum waldende, weorðan þyrfen.
Gif þu þæt angin fremest, idesa seo betste,
forhele ic incrum herran þæt me hearmes swa fela
Adam gespræc, eargra worda.'

'Eat this fruit! Then your eyes will afterwards become so clear that you might afterwards see over all the world and throne of the lord himself, and have his favour henceforth. You will afterwards rule Adam, if you control his will and he trusts in your words. If you say to him as a truth what example you yourself hold in your heart, because you carried out god's teaching, then he will give up in his heart the wicked opposition, the evil objection, if we two both speak to him skilfully. You persuade him eagerly that he should carry out your instruction lest you must become loathsome to god your ruler. If you accomplish that undertaking, best woman, I will conceal from your lord that Adam spoke to me so much insult, so many vile words.'

Here, the tempter creates a fictional world, with very few world-building elements; they are primarily function-advancing, unlike the final section, which contains no actions or events, but rather comprises a series of assertions and denials. The final section of the tempter's speech explains Adam's rationale for rejecting his entreaty, as well as a denial of Adam's claims and an assertion of genuineness (581–7):

The tempter's speech to Eve: part 4

'Tyhð me untryowða, cwyð þæt ic seo teonum georn,
gramum ambyhtsecg, nales godes engel.
Ac ic cann ealle swa geare engla gebyrdo,

heah heofona gehlidu; wæs seo hwil þæs lang
þæt ic geornlice gode þegnode
þurh holdne hyge, herran minum,
drihtne selfum; ne eom ic deofle gelic.'

'He accuses me of untruths, says that I am eager for injuries, a minister to the angry one, not at all an angel of god. But I know the whole race of angels so well, the high vault of heaven; the time was not so long ago that I eagerly served god my master, the lord himself with a faithful heart; I am not like a devil.'

In these four segments of *Genesis B*, many imagined states of mind, imagined future scenarios and anticipated reactions are presented, and indeed dominate the text: the tempter describes God's future state of mind towards Eve in response to the tempter's communication of a message to him, as well as the reverse of that situation. Eve is called upon to imagine also Adam's future state of mind as well as her own response to being able to rule him. Moreover, the tempter urges her to visualise her success with God, with Adam and with herself and to imagine the heightened perception that the fruit will provide (a vision, which, as Davis argues, the tempter himself provisionally grants Eve, thereby making real the earlier imagined reality).[78] The emphasis is on thought, image and speech, rather than on scene-setting or events, and a conflation of the discourse world (between the tempter and Eve) and the text world, with its presentation of this encounter by the narrator to the reader. This is an extreme example of the reader's reliance upon characters 'to provide the world-building and function-advancing information necessary to construct a coherent mental representation' of the text in Gavins's formulation.[79] These imaginings, reported perceptions and hypothetical propositions are all sub-worlds that are accessible only by the character (Eve) and reported to the reader by the narrator.

One specific means of creating a sub-world is through the use of negation. As Werth argues, negation draws attention to what it denies as it 'introduces not only the denial of the expectation, but also the expectation itself',[80] an idea that is particularly useful in making sense of the rhetorical power of the end of the devil's speech to Eve, in part 4, especially the emphatic conclusion, 'I am not like a devil' (587b), which draws attention to the precise opposite of its propositional content, for the reader, if not for Eve.

A further idea that is useful here is Palmer's concept of *intermentality*, or of one mind moving into another and, in the case of fiction, of creating the minds and mental processes of fictional characters, because narrative is essentially 'the presentation of fictional mental functioning'.[81] Our familiarity with this technique of mental visitation or impersonation in literature makes this process possible. To understand the text, the reader has to impersonate

78 Davis, 'Changing Senses in *Genesis B*'.

79 Joanna Gavins, '"Too Much Blague?": An Exploration of the Text Worlds of Donald Barthelme's *Snow White*', in *Cognitive Poetics*, ed. Gavins and Steen, pp. 129–44, at p. 139.

80 Werth, *Text Worlds*, p. 255.

81 Alan Palmer, *Fictional Minds* (Lincoln, NE, and London, 2004), p. 5.

Eve to some extent: share her vision through creating one's own mental representation of the imagined future benefits and punishments mentioned by the tempter, yet to feel in possession of superior knowledge. In *Genesis B*, the reader has access to the tempter, who assumes access to God's, Adam's and Eve's mental realms. The reader makes a mental construction of a reported mental construction. The tempter is thereby made prominent in that he 'filters the world switches through his own consciousness'.[82] He plays on Eve's mind by encouraging her to do likewise: construct a representation of God's reaction to her imagined action (558b–559b) – in fact, to impersonate God. The reader, in making sense of the text, is called upon to do likewise.

Genesis B is additionally interesting within the Text World framework in its thematisation of vision, and its use of the *megametaphor* of the fruit of knowledge. Werth explains that metaphor provides a double layer of text worlds through the act of 'opening up an area of experience in terms of which the discourse topic can be (partially) interpreted'.[83] This 'double-vision' modality allows movement back and forth between the two frames or components of the metaphor, in that it allows 'target domain meanings to be understood in terms of source domain properties'.[84] Werth's own figure of metaphor itself as double vision takes on a further layer of meaning when applied to actual metaphors of vision in *Genesis B*. In this text, the widespread conceptual metaphors of knowledge as vision and thinking as movement combine with specific individual metaphors (e.g. thought is a seething liquid; resolution is a fragile object) to create a complex network of associations between perception and human choice. Here, the megametaphor prevails: knowledge is vision but also, paradoxically, blindness and delusion. Similarly, strength is obedience is blindness; choice is death is fruit; and weakness is mobility is manipulation. A Text World approach provides a way of analysing this complex systematically.

The significance of the scriptural antecedent and actual diction of the text can also be illuminated within the Text World framework. We can see textual imposition at work, where the reader's interpretation is guided by loaded terms.[85] For example, epithets for Satan in *Genesis B* make his villain status clear: 'wraðmod' (547a), 'one angry in mind'; 'laðan' (601a), 'hostile one'; 'sceaða' (606b), 'devil'; and 'forhatena' (609b), 'devil'. Similarly, descriptions of the fruit are loaded: it is called a 'weorcsumne wæstm' (594a), 'noxious fruit'. Eve is repeatedly called 'good' by the devil, in open flattery and at odds with her behaviour: 'Eue seo gode' (612a), 'Eve the good'; and 'idesea seo betse' (578b), 'best woman'; and beautiful by the narrator: 'sceone gesceapene' (549a), 'beautiful creature', who also accounts for her susceptibility by ascribing to her a divinely assigned weaker will: 'hæfde hire wacran hige / metod gemearcod' (590b–591a), 'God had assigned to her a weaker disposition'. Her culpability is

[82] Stockwell, *Texture*, p. 111.

[83] Paul Werth, 'Extended Metaphor – A Text-World Account', *Language and Literature* 3.2 (1994), 79–103, at p. 95.

[84] Ibid., p. 96.

[85] Stockwell, *Texture*, p. 166.

expressed through the next line: 'heo hire mod ongan / lætan æfter þam larum' (591b–592a), 'so that she began to allow her mind (to follow) those counsels'.

Readers are guided by these motivated lexical choices to collect and deploy this information in *Genesis B*, and respond to it cognitively in very different ways. The difference can be accounted for in part by the Theory of Text Worlds, where some readers see the sub-worlds as creations of the narrator, others see them as created by the tempter, others by Eve. In a text that focuses so extensively on perception, imagination and delusional vision, this slippage among sources of information allows a particularly wide range of readerly creativity in synthesis and interpretation, and in turn enacts the crossing of that boundary that the text explicitly treats.

World-switching in Wulf and Eadwacer and the desire for cohesion

Well beyond the differences of opinion regarding the interpretation of *Beowulf* and *Genesis B*, grappling with the meaning of the short elegiac poem *Wulf and Eadwacer* has become a form of ultimate challenge for Anglo-Saxonists. Critics agree on little regarding this poem, except on the difficulty of creating narrative cohesion from the collection of allusive remarks that constitute it. As a basis for the discussion on what Text World Theory can bring to this challenge, the whole text is provided below, along with one possible translation.

Leodum is minum   swylce him mon lac gife;
willað hy hine aþecgan,   gif he on þreat cymeð.
Ungelic is us.
Wulf is on iege,   ic on oþerre.
Fæst is þæt eglond,   fenne biworpen.
Sindon wælreowe   weras þær on ige;
willað hy hine aþecgan,   gif he on þreat cymeð.
Ungelic is us.
Wulfes ic mines widlastum   wenum dogode;
þonne hit wæs renig weder   ond ic reotugu sæt,
þonne mec se beaducafa   bogum bilegde,
wæs me wyn to þon,   wæs me hwæþre eac lað.
Wulf, min Wulf,   wena me þine
seoce gedydon,   þine seldcymas,
murnende mod,   nales meteliste.
Gehyrest þu, Eadwacer?   Uncerne earne hwelp
bireð wulf to wuda.
Þæt mon eaþe tosliteð   þætte næfre gesomnad wæs,
uncer giedd geador.

It is as if someone gave my people a gift; they plan to kill him if he comes into their troop. It is different for us. Wulf is on an island, I on another. That island is fast bound with fens. There are murderous men on the island; they plan to kill him if he comes into their troop. It is different for us. I endured with hope Wulf's long journey. When it was rainy weather and I sat lamenting, then the battle-bold one embraced me with his arms. There was joy for me in that, yet there was also loathing. Wulf, my Wulf,

> hopes for you have made me sick, your rare visits, my mourning mind, not at all lack of food. Do you hear Eadwacer? A wolf bears our wretched cub to the woods. One may easily separate that which was never united, our tale together.

It is unsurprising that many scholars have considered *Wulf and Eadwacer* to be one of the most intriguing and enigmatic of extant Old English poems. Its allusive addressees and ambiguous relationship references make it unclear to the modern audience who is saying what to whom and why, notwithstanding the straightforward syntax of this text and its repetitive phrasing. The riddling qualities of this text noted by many critics[86] have provoked many creative responses. Readers have succumbed to the desire for narrative clarity or at least a reasonable solution in this textual encounter by suggesting several different scenarios, mostly based on imaginative guesswork, and in turn, often attracting harsh criticism from other scholars who reconstruct the pieces and supplement them differently. It is common for critics to profess to being unconvinced by another's inference-making. As Henk Aertsen says, 'any interpretation must necessarily be based on a number of assumptions and suppositions which are impossible to verify', before asserting his own scenario, for which 'there is, I admit, no hard evidence to support this reading'.[87] Most other critics do likewise, setting out to create a narrative arc that accounts for the words and ideas of the text, and being entirely willing to fill in the large gaps themselves. The readiness of critics to go out on a limb to this extent attests the strong drive for resolution, as well as the predominantly narrative approach to the search for meaning in this text.[88]

Proposed readings of this poem are based on assumed common ground – the likelihood of the certain typical human interactions underlying the textual pronouncements – and are predicated on an assumption of consistent human emotional and rational experience over time (i.e. assume consistent evolved human cognitive and emotional functioning), though of course, what seems most likely to one individual might seem far-fetched to another. The bulk of critics have favoured a love triangle involving a woman, her lover Wulf the outlaw, and unloved husband, Eadwacer, a scenario first proposed by the earliest modern interpreters and editors of the poem. Peter S. Baker

86 Henk Aertsen reviews the scholarship in this area, and convincingly demonstrates the unlikelihood of the riddle interpretation of the poem, notwithstanding some riddling features. See '*Wulf and Eadwacer*: A Woman's *cri de coeur* - For Whom, For What?', in *Companion to Old English Poetry*, ed. Henk Aertsen and Rolf H. Bremmer Jr (Amsterdam, 1994), pp. 119–44, esp. pp. 125–9. John D. Niles agrees that this text 'is meant to be deciphered on the basis of hints and allusions', *Old English Enigmatic Poems and the Play of the Texts* (Turnhout, 2006), p. 46. In this study of riddles and riddle-like poetry, Niles refers to the short *Exeter Book* lyrics, including *Wulf and Eadwacer*, as '"talking points" rather than as texts whose meaning sits upon them like a label' (p. 309).

87 Henk Aertsen, 'Second Thoughts on *Wulf and Eadwacer*', pp. 3 and 6 (an expanded online version of a 2001 conference paper: www.henkaertsen.nl/W&E2ndthoughts.pdf), following up some lines of thought in his earlier article, noted above, n. 86.

88 A notable exception is Alain Renoir, '*Wulf and Eadwacer*: A Non-Interpretation', in *Medieval and Linguistic Studies in Honor of Francis Peabody Magoun, Jr*, ed. J. B. Bessinger Jr and R. P. Creed (New York, 1965), pp. 147–63, where the elegiac rather than narrative coherence of the text is asserted: 'the poem stands apart from its meaning', at p. 160.

sums up the first century of scholarship on this poem as confirming this early scenario, which Baker himself modifies only slightly (so that Wulf is her husband, Eadwacer her jailer).[89] Baker, like Davidson,[90] ascribes the poem's ambiguity to poetic mastery rather than entirely to our lack of access to culturally specific information that would fill in some of the huge gaps in the text's narrative. This ambiguity is sufficient for more than one scholar to read the references entirely differently, so that the poem communicates a mother's lament for a lost child.[91] Indeed, one of these scholars, Marijane Osborne, claims that interpretation has been skewed by 'our own romantic preference for a story about lovers',[92] a statement that both implicates our tendency to interpret the facts and fill in the textual gaps according to a privileged narrative and emotional template, and also silently acknowledges culturally embedded text world making. Both assumptions can be explored further using a Text World framework.

Similarly, the contents of the common ground and readerly inferencing can be examined further. Many textual interpretations are more emphatically informed by known Anglo-Saxon or Germanic cultural practices or knowledge, or in turn deploy the text as evidence for Anglo-Saxon views on women.[93] The word 'wulf', which occurs five times in the poem, has invited a particularly large amount of scholarly attention and vigorous debate centred on the possibility that the connotation 'outlaw' is appropriate in this context. Eric Stanley has roundly debunked this idea,[94] though it persists in recent scholarship, and has been linked with the idea that a werewolf is invoked here.[95] Other interpretations similarly rely on information encoded in single words, so seem either to extrapolate beyond being convincing, or are content to provide a tentative interpretation, or rather point to the culturally sensitive nature of poetic meaning and our lack of familiarity with a particular narrative invoked. At any rate, the poem has certainly attracted more than its share of energetic interpreters determined to resolve or at least account for its manifest ambiguity and allusiveness, with the result that any scholarly interpretation must wittingly or not derive from earlier readings which have fallen into one dominant and one minor track (i.e., love triangle and mother's

89 Peter S. Baker, 'The Ambiguity of *Wulf and Eadwacer*', *Studies in Philology* 78.5 (1981), 39–51, p. 50.

90 Arnold E. Davidson, 'Interpreting *Wulf and Eadwacer*', *Annuale Mediaevale* 16 (1975), 24–32.

91 Marijane Osborne, 'The Text and Context of *Wulf and Eadwacer*', in *the Old English Elegies: New Essays in Criticism and Research* , ed. Martin Green (London and Toronto, 1983), pp. 174–89; Dolores Frese, '*Wulf and Eadwacer*: The Adulterous Woman Reconsidered', in *New Readings on Women in Old English Literature*, ed. Helen Damico and Alexandra Hennessey Olsen (Bloomington, 1990), pp. 273–91; and J. A. Tasioulas, 'The Mother's Lament: "Wulf and Eadwacer" Reconsidered', *Medium Ævum* 65 (1996), 1–18.

92 Osborne, '*Wulf and Eadwacer*', p. 184

93 For the latter, see Helene Scheck, 'Seductive Voices: Rethinking Female Subjectivities in *The Wife's Lament* and *Wulf and Eadwacer*', *Literature Compass* 5.2 (2008), 220–7.

94 Eric Stanley, 'Wulf, My Wolf!', in *Old English and New: Studies in Language and Linguistics in Honor of Frederic G. Cassidy*, ed. Joan H. Hall, Nick Doane and Dick Ringler (New York: Garland, 1992), pp. 46–62.

95 Sonja Daniëlli, '*Wulf, Min Wulf*: An Eclectic Analysis of the Wolf-Man', *Neophilologus* 90 (2006), 135–54. Notwithstanding their similar titles, this article does not cite Stanley's.

loss of a child respectively). Indeed, more than one critic regards the text as so overinterpreted as to prejudice any modern-day access to its textual rather than historical meaning,[96] the result being a severe limiting of Jauss's 'horizon of expectations'.[97]

This metainterpretation has arisen because the relationships among its referents of this stubbornly difficult poem are difficult for modern readers to conceptualise. At this remove, it is unlikely that any new reading or theoretical framework will resolve the residual riddles or remedy the obliqueness of the narrative of this text, and there is no claim that Text World Theory can accomplish what other paradigms and approaches have failed to do. As a discourse framework, however, Text World Theory can offer some insight into the affective quality of the text and the cognitive sources of its interpretive conundra, and in turn *Wulf and Eadwacer* can provide a robust sample text with which to test some of the claimed benefits of this theory. For example, by asking what possible worlds the Anglo-Saxon audience would have entertained in order to makes sense of this text (especially as legendary material is perhaps involved), we might estimate what degree of ambiguity is intended or inevitable, and what stems from our inability to create narrative coherence at this remove. Even more interestingly, an analysis of those possible worlds requires us to consider how we approach the emotional charge of the text, and to address the particular assumptions we bring to it about common ground, the emotional life of humans,[98] and the affective potential of poetry. Our creation of worlds and plot scenarios for this text are predicated on our belief in the stability of human emotional evolution over time and cultures, and the sameness of human (and gendered) reactions to love, separation, relationships with lovers, husbands, those who force their physical attentions on us, and to children. Attention to that process in turn sheds light on textual dynamics, even when meaning is allusive.

The mental representations we modern readers create from the textual clues of *Wulf and Eadwacer* arise from a slippery text world and multiple world shifts within the text, enacted with minimal or apparently non-existent markers. The reader is required to shift attention in time and space without very clear or prominent deictic or referential indicators. We are required to manage and negotiate several provisional and incompatible scenarios at once, and perceive the relevance of several generalised, gnomic utterances and repeated remarks, with no promise or delivery of resolution, or even further

96 For a summary of recent views, and a deconstruction of any notion of a stable meaning available from the text, see James J. Donahue, '"Of this I Can Make No Sense": *Wulf and Eadwacer* and the Destabilization of Meaning', *Medieval Forum*, ed. George W. Tuma and Dina Hazell, 2004, accessed 5 September 2011 at: http://www.sfsu.edu/~medieval/Volume4/Donahue.html .

97 Hans Robert Jauss, *Toward an Aesthetic of Reception*, trans. Timothy Bahti, introd. by Paul de Man (Brighton, 1982), p. 23.

98 Or dogs, in the case of W. J. Sedgefield, 'Old English Notes', *Modern Language Review* 26 (1931), 74–5: 'a female dog of a romantic temperament is dreaming' (p. 74). Most readers have not been convinced by the likelihood of this mental representation, which is not considered here, though a variation, that the poem refers to wolves, is argued by Peter Orton, 'An Approach to *Wulf and Eadwacer*', *Proceedings of the Royal Irish Academy* 85C (1985), 225–58.

specificity at the end of the text. The uncertain status of the participants – the female speaker, indicated by the feminine endings of the adjectives 'reotugu' (10) and 'seoce' (14); Wulf, Eadwacer, and the whelp – makes the discourse situation within the text even more complicated. It is unclear, for instance, who the primary addressee is, and whether 'wulf' is a name, a reference to the animal, a symbolic indicator, or an epithet.

But space and time are emphasised within the text by virtue of the prominence of those deictic markers that do exist, especially prepositions, and the textual focus on separation. In Stockwell's terms, separation – physical, temporal, perceptual and metaphysical, – becomes an attractor,[99] sustained by the reiteration of words denoting difference ('ungelic') and isolation on an island in this poem. This discourse technique is complicated by apparent movement from the general state of affairs to the specific situation, from the past to the present, and the lack of clear indication of when those movements occur. This process culminates in the speaker and her current situation.

If we look beyond the precise though allusive wording of this text to its discourse situation, the entire poem appears to be a form of mental representation or memory itself – an impassioned monologue from a woman expressing her grief at being separated from her loved one (or ones) – but this discourse world is implied grammatically rather than otherwise, so source tagging and metarepresentation are weak. On the other hand, the underlying affective content is strong: no reader, to my knowledge, has understood that the speaker expresses an unconcerned or positive attitude to the situation. It comprises memories of the past, distress at the present, fears and hopes for the future. The reader is required to hold these three worlds, and the multiple discourse worlds of the woman and her interlocutors, in mind throughout the poem, and to accommodate multiple rapid world switches. A line-by-line chart will indicate the world switches required to track the time and place references.

World Switches in Wulf and Eadwacer

[unknown addressee(s)]

- Line 1 – present time: world of speaker's people
- Line 2 – switch to future time: world of speaker's people and 'he'
- Lines 3 and 8 – switch back to present time: unspecified world, including speaker
- Line 4 – present time: world of Wulf's island and switch to speaker's separate island (main discourse world of poem, and specification of line 3 world [with ambiguity])
- Lines 5–7 – present time: switch to world of Wulf's island (clarification of world of line 2)
- Line 9 – world switch to past time: speaker's place unspecified (?island)
- Line 10–12 – still past time, but switch to a particular point: speaker's place unspecified (?island)

[99] Stockwell, *Cognitive Poetics*, pp. 42 and 98.

- Lines 13–15 – world switch to present time, with reference to future and future in the past [hopes of you]: speaker's island
- Line 16a – world switch to new addressee, present time: speaker's island
- Line 16b–17 – world switch to past time: place of event – unspecified wood
- Line 18–19 – world switch to unspecified time and place, with past and continuous reference ('that never was').

The overwhelming sensation is of movement of thought, flitting around in time and space: a movement of figure and ground, a shifting of attractors, a movement of world edges as switches occur. As Stockwell argues, 'it is the crossing of world-boundaries itself that generates this sense of texture';[100] 'texture is in the edges',[101] because objects have both shape and directionality, and is a generic feature of lament poems, especially useful for the creation of empathy. This poem is almost entirely edges – the movement is as prominent as the content and can be seen to carry a good deal of its emotional charge.

Some of the switches concern addressivity – the speaker seems to be talking to several different people at once, including herself, and to the world in general, thereby creating a dynamic reading. This 'readerly sensation of transportation'[102] can be analysed with Text World Theory as 'a form of projection of identity or at least a close comparison of an alternate state of affairs',[103] or an 'interrelated set of belief worlds that the reader is invited to imagine'.[104] The reader is invited to live in the confusion of unsettled world creation. The image schemas here are binary: of here and there, in and out, joy and sadness. The affective quality of this text arises from these techniques, which express a lack of certainty and a denial of full sense of belonging, notwithstanding definite placement in time and space. The reader has to make a considerable conceptual and emotional investment in decoding the text via the creation of several possible and incomplete text worlds, creating an 'emotional feedback loop'.[105]

The reader, even one contemporary with the text, has to negotiate this moving terrain and time switches, and the high degree of vagueness in the text, in terms of both time and place. It is clear that the story is not all there, and so a known narrative, text type, or proximate intertext was required to fill in the large gaps left by the text and its multiple world shifts.[106] On the other hand, perhaps the desire for narrative coherence has tempted modern readers to privilege the search for conceptual cohesion over other textual considerations: perhaps the shifts are central to the text's communication of distracted grief and a generalised rather than specific evocation of human sympathy, or the literary effect relies on ambiguity in order to foreground affective involvement

[100] Stockwell, *Texture*, p. 133.
[101] Ibid. p. 107.
[102] Stockwell, *Texture*, p. 91.
[103] Ibid.
[104] Ibid., p. 98.
[105] Ibid., pp. 95 and 98.
[106] Another possibility is that the extant text is fragmentary or incomplete, but, in the absence of any physical or other evidence to support this idea, most critics do not pursue it.

rather than narrative meaning. This text insists that the reader make the major contribution to meaning, in terms of building entire worlds from brief, allusive comments. It structures emotional rather than narrative coherence.

The conceptual structure of *Wulf and Eadwacer* can be disclosed in part by an analysis of its world-building structures, but also by its affective and psychological points of focus. The Text World approach can therefore be usefully combined with more strictly lexical and stylistic approaches to the analysis of poetic meaning. For instance, here a dominant first-person perspective with dual-person reference is encoded lexically. The 19-line poem contains 117 words, of which 9 are first-person pronouns or possessive adjectives, 4 are dual/plural and 2 are second-person, so about 13 per cent of the text marks 'I–you–we'. Other semantic preoccupations are 'wulf' (5 times) and references to the mental life ('aþecgan' (twice), 'intend'; 'wenum', 'wena' 'hope'; 'mod', 'mind'; and perhaps 'hogode' if we accept the emendation of 'dogode' on 9b[107]), perception ('gehyrest', 'hear'), emotion ('reotugu', 'weeping'; 'wyn', 'joy'; 'lað', 'loathing'; 'murnende', 'mourning'; and 'earne', 'wretched'). Taken together, these points of focus construct a distinctive texture of an emotional relationship between two people, brought to a critical culmination in the final line where the paradox of joined separability is expressed along with a metafictive reference to 'our giedd geador', 'our tale together' which has just been reflected on in the poem. The centrality and persistent ambiguity of 'wulf' complicates this texture, but does not disturb the tone of melancholy distress, and the expression of unnatural and painful separation that pervades the text. The lovers are divided physically, and the woman is even divided from herself: 'wæs me wyn to þon,   wæs me hwæþre eac lað', 'there was joy for me in that, yet there was also loathing'. The very discourse situation of the text, with its multiple implied recipients of Wulf (13a), Eadwacer (16a), herself and the world at large (gnomic utterances, such as lines 18–19), fragments the message and complicates the act of communication.

The texture of this poem is triangulated by sadness, separation and the multiple ideas clustered around 'wulf'.[108] In Stockwell's terms, the irrelevant background information is suppressed in order to 'enhance the figural attractors in their texts'[109] to create a particular sort of literary richness or resonance that invites empathetic engagement with the reader that is an embodied affective response.[110] This idea relies on the principle of embodiment (discussed above), which assumes a direct connection between physical experience of being in the world and cognitive activity, here specified as 'the physicality of literary reading',[111] a concept that takes on even greater relevance to oral

107 For retention of the MS reading, 'dogode', a hapax presumed to carry animal connotations ('tracked'), see Marijane Osborne, '*Dogode* in *Wulf and Eadwacer* and King Alfred's Hunting Metaphors', *American Notes and Queries* 13 (2000), 3–9.

108 My argument at this point is similar to that of Renoir, '*Wulf and Eadwacer*: A Non-Interpretation', who refers to four main themes in the poem: separation, union, hostility and suffering (p. 152), though it is framed differently.

109 Stockwell, *Texture*, p. 29.

110 Ibid., p. 56.

111 Ibid., p. 58.

circumstances of composition and dissemination of texts as in Anglo-Saxon England and the alliterative, metrical force of poetry. This information indicates the intensely affective fabric of the possible worlds the Anglo-Saxon audience would have created in order to makes sense of this text, and the narrative ambiguity used as a vehicle for emotional power specifically and the affective potential of poetry generally. The mental representation the text triggers is an affective rather than a narrative one.

The richness of *Wulf and Eadwacer* – its emotional intensity – is the result of the reader's contribution of emotional disposition, experience and involvement. This emotional investment is defined by Stockwell as 'the accompanying feeling of literary resonance'.[112] Readers are up to this task of responding to the demands placed on them precisely because the text drives inference-making and the process of making coherence from its propositions and the mental representations it gives rise to.

Building on text-world theory in Anglo-Saxon studies

As I hope these textual examples suggest, Text World Theory provides new perspectives on Anglo-Saxon texts that can support and combine with other cognitive approaches, especially Conceptual Blending Theory, as well as more traditional approaches such as lexical, stylistic and literary theoretical ones. Chief contributions of this approach include the way it presents coherence as structured, and the way it emphasises how the text itself and the propositions that can be worked out from it act in concert in any act of interpretation. This approach is more prominently linguistic that others, but is broadly very useful, as its central ideas can be adapted beyond linguistic analysis. Most importantly, its central focus on the generation and management of imaginative scenarios during the act of comprehension is a robust and exciting idea that is worth consideration by a wide range of literary scholars. Moreover, a theory that accommodates how textual encounters generate mental representations is particularly useful for application to a textual corpus such as the Old English one that features so pronounced an interest in imagination, perception and the inner life generally. This thematic preoccupation with the conceptual building-blocks of the Text World paradigm invites a two-fold consideration of how imagination works in vernacular poetry: both within and in response to the text.

Text World Theory provides an approach to conceptualisation that allows the psychological processing of language to be tracked to show how prominence is driven through the force dynamics of texts. It insists that assumptions governing any act of meaning-making are collaborative, provisional and reliant upon a negotiated and evolving set of propositions that are blended into a coherent system within a conceptual space. These ideas, as well as those described above, provide a new set of concepts that can usefully be

[112] Ibid., p. 63.

combined with well-established interpretive frameworks in Anglo-Saxon Literary Studies, and also with Cognitive Cultural Studies, treated in the next chapter.

5

Cognitive Cultural Studies

An emerging set of approaches

'Cognitive Cultural Studies' refers to an emerging interdisciplinary group of approaches, which is developing into one of the most interesting and dynamic areas of inquiry relevant to Literary Studies today. At its core is an interest in the interplay between brain and environment – cognitive architecture and cultural forces – a focus that requires a genuine interdisciplinary crossing of the arts/science divide that has been a desideratum in Psychology and Anthropology for at least two decades.[1] John Tooby and Leda Cosmides laid out the argument very clearly in 1992 in their integrated model of human social and cognitive functioning:[2]

> The rich complexity of each individual is produced by a cognitive architecture embodied in a physiological system, which interacts with the social and non-social world that surrounds it. Thus humans, like every other natural system, are embedded in the contingencies of a larger principled history, and explaining any particular fact about them requires the joint analysis of the principles and contingencies involved.

During the last two decades, developments and cross-disciplinary research into this complex interaction of culture and cognition from a range of social and hard sciences have provided the context for a more nuanced consideration of the 'culture' side of the issue, using the tools and concepts developed in the humanities. This mutual impact of social and mental functioning plays out in several different forms of scholarly inquiry, and has gained momentum in the last few years particularly, in the emergence of the scholarly collective, 'Cognitive Cultural Studies'.

The range and focal interests of scholars working in this field are conveniently laid out in a recent volume edited by Lisa Zunshine, *Introduction to Cognitive Cultural Studies*.[3] This book contains fourteen essays that deal with some of the many aspects of how culture and cognition interact, and lays the groundwork for combining the cognitive science of meaning-making with an informed understanding of literature as cultural representation. The four parts of this book reflect the major trends in this field: 'Literary Universals'

1 John Tooby and Leda Cosmides, 'The Psychological Foundations of Culture', in *The Adapted Mind: Evolutional Psychology and the Generation of Culture*, ed. Jerome H. Barkow, Leda Cosmides and John Tooby (Oxford, 1992), pp. 19–136.

2 Ibid., p. 21.

3 Lisa Zunshine, ed., *Introduction to Cognitive Cultural Studies* (Baltimore, 2010).

(the study of widely found types of literary techniques and their basis in embodied experience); 'Cognitive Historicism' (the study of the interaction between cognitively similar human minds and their variable historical circumstances); 'Cognitive Narratology' (the impact of mental processes on the practices of telling and interpreting stories); and the self-explanatory 'Cognitive Approaches in Dialogue with other Approaches (Postcolonial Studies, Ecocriticism, Aesthetics, Postructuralism)'. These sections show the wide applicability of cognitive approaches, and particularly their capacity to combine with other established schools of thought and methodologies.

This field has grown out of Cognitive Literary Studies, earlier defined by F. Elizabeth Hart as 'an interest in the cognitive, from a literary perspective, is an interest in exploring how both the architecture and the contents of the human brain/mind – both in terms of its online processing of information and its evolutionary history – may contribute structurally to the writing, reading, and interpretation of texts'.[4]

The substitution of the word 'culture' for 'literature' in the new field, Cognitive Cultural Studies, reflects the broadening of scope to cultural practices beyond the literary, and the acknowledgement of the enmeshed interaction of literary texts, cultural contexts and cognitive processes occurring within an embodied mind. At its essence is an interest in the complex mutual impact of biology and context, in 'relationships between the humanly universal and the culturally specific'.[5] As Alan Richardson explains, one key aim is to show precisely how 'cultural and historical differences will emerge more clearly and cleanly when set against what appear to be stable and invariant aspects of human cognition and behavior'.[6]

Despite this very broad remit, this approach avowedly resists sustained interaction with many of the frameworks treated in the chapters of the present study above – Cognitive Poetics, conceptual blending or 'empirical studies of literature'[7] – even though they all rely to varying degrees on an acknowledgement of embodied cognition and its linguistic expression, especially as conceived by Lakoff and Johnson (discussed above). Rather, this approach explores the constitutive role of cognition in the creation of culture – how cultural practices emerge as a result of mental processes, rather than vice versa. It brings the science of the evolved human mind to bear on the conceptual

4 F. Elizabeth Hart, 'The Epistemology of Cognitive Literary Studies', *Philosophy and Literature* 25 (2001), 314–34, at p. 319.

5 Zunshine, *Introduction*, p. 2.

6 Alan Richardson, 'Facial Expression Theory from Romanticism to the Present', in *Introduction*, ed. Zunshine, pp. 65–83, at p. 68.

7 Zunshine, *Introduction*, p. 4. This disinclination for cross-fertilization between the fields of Cognitive Poetics and Cognitive Cultural Studies is apparently an unfortunate by-product of the persistent disciplinary boundary between (and different conference attendance patterns of) Linguistics and Discourse Analysis on the one hand and Literary and Cultural Studies on the other, though there is some evidence that the boundaries are breaking down: see, for example the articles in David Herman, ed., *The Emergence of Mind: Representations of Consciousness in Narrative Discourse in English* (Lincoln, NE, 2010). This book charts the different representations of consciousness in English literary discourses from the Old English period to the present day.

frameworks of Cultural Studies as envisaged by its founder, Raymond Williams, and developed by later interpreters.[8] Zunshine's detailed engagement with Williams's framework makes a strong case for the compatibility of Cognitive Cultural Studies with, and natural expansion of, that framework.

In particular, cognitive cultural critics investigate the relationship between brain and culture, between the embodied mind and the historicised moment, using the tools of science. As Zunshine explains:[9]

> To understand why a certain new interpretation of 'sensory information', say a movie about Prohibition-era gangsters made in the 1990s, becomes, or does not become, particularly appealing (or, as Williams would say, communicable), we have to situate it in its rich historical context. We have to ask what specific challenges of this moment – economic, theological, sexual, political, emotional, linguistic, ecological, personal (and so forth: ways of framing these challenges are infinite) – this way of seeing the world engages. But at the same time we also have to ask how it engages them so as to be relevant (again: communicable) to specifically human cognition; for instance how it interacts with our theory of mind, with our tendency to engage in analogical thinking, with our embodied perception of the world, with particularities of our emotional response to representations and how it forges (or severs) connections between our sensory modules.

As this comment suggests, human engagement and interest relies on the discernment, selection and prioritisation of information, an idea in keeping with the central focus of Cognitive Poetics on the motivation selection of specific material from the total store, and its clustering in response to textual clues in the amalgamation of meaning. Both schools of thought foreground the importance of motivated inferencing, and stress the cognitive complexity of that process, whereby new information is generated via analogical processes of reasoning.

But Cognitive Cultural Studies diverges from Cognitive Poetics in its explicit deployment of the concepts and methodologies from Cognitive Science, especially Neuroscience and Evolutionary Psychology, to explore topics such as Theory of Mind, or mind-reading; distributed or social cognition; facial recognition and human visual information processing more broadly; analogical reasoning; and the neural underpinnings of imagery based on bodily sensations. Cognitive Science provides insight into the neural mechanics of representation, modelling and the creation and storage of imagery in the brain. In turn, Cognitive Cultural Studies supplies to Cognitive Science detailed and nuanced insight into literature as a product of – indeed a manifestation of – both cognitive processes and embedded cultural practices.

More specifically, in the Cognitive Cultural framework, literature is treated as not only a product of cognitive adaptations, but also a facilitator of the energetic or extreme use of those adaptations. Engagement with literary texts, the argument goes, is possible only because the human brain has evolved with

8 Raymond Williams, *The Long Revolution* (Harmondsworth, 1965), p. 40, cited in *An Introduction*, ed. Zunshine, p. 8.

9 Zunshine, *Introduction*, pp. 15–16.

the capacity for imagination, perception, knowledge, memory and language, and literary texts in turn sharpen those skills by requiring their application in concentrated or focused form. For instance, Monica Fludernik points out the benefit of experientiality supplied by fiction, when she argues that narrative provides us with an opportunity to experience another consciousness.[10] In turn, that experience with texts allows the human mind the opportunity of experimenting with trying on different consciousnesses, and fine-tuning cognitive abilities that are able to be deployed elsewhere in human experience. Zunshine takes this line when she argues that we enjoy flexing our cognitive abilities, including mind-reading and metarepresentation (discussed further below, pp. 109–10), via engagement with fiction.[11] One of the grounding assumptions of Cognitive Cultural Studies is that these mental processes are influenced, even constrained, by cultural forces. Ellen Spolsky has shown how this idea plays out in Early Modern Studies.[12] Like other scholars working in this field, Spolsky incorporates scientific developments into cultural history and literary production, as well as exploring the cognitive underpinnings of particular cultural representations. Similarly, Alan Palmer has shown very fully how much collaborative cognitive activity is reported in literary texts in his study of the minds of characters in novels, most recently in his work on social minds represented in literature: how decisions and opinions are arrived at collectively rather than individually, where a particular group of people can be said to hold a certain view.[13] In his 'externalist' perspective, he considers those aspects of the mental life that are public or social. He argues that collective thinking is widely operative in the fictional worlds of novels, where a vigorous interaction between individual and social minds underpins the narrative. Palmer argues throughout *Social Minds* that novels, of which *Middlemarch* is an outstanding example, present extensive shared or group cognition. He concentrates on nineteenth-century novels, though concedes that his method could be applied to more recent ones,[14] without apparently considering that it might apply to earlier texts too. It might be argued, though, that this technique of representing distributed cognition representation is not the sole property of the modern novel: see for example *Elene* (894–898a):

> Ða wæs in þam folce   on ferhðsefan,
> ingemynde,   swa him a scyle,
> wundor þa þe worhte   weoroda dryhten
> to feorhnere   fira cynne,
> lifes lattiow.

10 Monica Fludernik, *Towards a 'Natural' Narratology* (London, 1996), and updated in Fludernik, 'Natural Narratology and Cognitive Parameters', in *Narrative Theory and the Cognitive Sciences*, ed. David Herman (Stanford, 2003), pp. 243–67, esp. pp. 245–6.

11 Lisa Zunshine, *Why We Read Fiction: Theory of Mind and the Novel* (Columbus, 2006), p. 25.

12 Ellen Spolsky, *Satisfying Skepticism: Embodied Knowledge in the Early Modern World* (Aldershot, 2001), p. 4.

13 Alan Palmer, *Social Minds in the Novel* (Columbus, 2010). See also Palmer, 'Storyworlds and Groups', in *Introduction*, ed. Zunshine, pp. 176–92, esp. pp. 184–6.

14 Palmer, *Social Minds*, p. 37.

> Then there was remembrance in the mind of that people, as it ever should be to them, those wonders that the lord of hosts, the ruler of life wrought as salvation to humankind.

Here, remembrance is presented as distributed among the whole population, or produced socially, which is precisely the phenomenon described by Palmer.

Palmer's argument relies on recent scholarship in Cognitive Sciences on distributed or situated cognition, where the interrelationship of culture and cognition is advanced as a productive lens through which to understand the human mind. Edwin Hutchins advanced the benefits of such interdisciplinary study of cognition as a cultural process in his influential book, *Cognition in the Wild*, in 1995.[15] In this book, Hutchins argues that his example, open-sea navigation, is a cognitive process that can be understood only within a larger system that includes culture, as a broader, collective context in which cognitive functions are enacted: as he claims, 'cognition is a fundamentally cultural process'.[16] A similar idea was presented by Andy Clark and David J. Chalmers in their flagship article, 'The Extended Mind', in 1998, where they asked, provocatively, 'where does the mind stop and the rest of the world begin?',[17] a question that many critics have responded to in the intervening period.[18] The argument of Clark and Chalmers for 'active externalism' has opened up Cognitive Science to even greater interdisciplinarity than it had previously enjoyed, in order to consider more fully the role of culture in cognition.

More recently, Clark has explained the imperative for a broader base for this research because 'cognition leaks out into the body and the world',[19] so that broader context requires consideration and the recruitment of a wide range of academic specialisations. This view extends the mind into the environment, specifically considering it to emerge 'at the productive interface of brain, body, and the social and material world',[20] a view that many scholars have taken up in various ways.[21] One scholar in this area who takes a diachronic approach is John Sutton, who champions the idea of 'Historical Cognitive Science' in order to examine the cultural basis of Cognitive Science with diachronic depth.[22] As Sutton argues, 'culture, artifice, and moral practice [are] inevitable, structuring supplements which construct and maintain the

15 Edwin Hutchins, *Cognition in the Wild* (Cambridge, MA, 1995).

16 Ibid. p. 374.

17 Andy Clark and David J. Chalmers, 'The Extended Mind', *Analysis* 58 (1998), 10–23, at p. 10.

18 See the responsive articles in Richard Menary, ed., *The Extended Mind* (Cambridge, MA, and London, 2010), which also reprints the original article by Clark and Chalmers at pp. 27–42.

19 Andy Clark, *Supersizing the Mind: Embodiment, Action, and Cognitive Extension* (Oxford, 2011), p. xxviii.

20 Ibid. p. 219.

21 See, for instance, the variety of extended cognition argued for in Anthony Chemero, *Radical Embodied Cognitive Science* (Cambridge, MA, and London, 2009), where a dynamic system between an agent and his or her environment is proposed.

22 See especially John Sutton, 'Exograms and Interdisciplinarity: History, the Extended Mind and the Civilizing Process', in *The Extended Mind*, ed. Menary, pp. 189–225. See also Sutton's 'Body, Mind, and Order: Local Memory and the Control of Mental Representations in Medieval and Renaissance Sciences of Self', in *1543 and All That: Word and Image in the Proto-Scientific Revolution*, ed. G. Freeland and A. Corones (Dortrecht, 2000), pp. 117–50; and *Philosophy and Memory Traces: Descartes to Connectionism* (Cambridge, 1998).

biological processes that they simultaneously and deeply transform'.[23] Given the similarity of this proposition with the core claims of Cognitive Cultural Studies, one might expect to see in the future greater a convergence between this field and the sections of Cognitive Science that deal with distributed cognition, particularly the work of Clark, Chalmers, Hutchins and Sutton,[24] a convergence that the current study aims to advance.

That process has already begun in some quarters, and shows signs of genuinely new insights. One of the most interesting and fruitful lines of inquiry within this interdisciplinary dialogue over the negotiated boundary between the natural and the cultural, and one which focuses specifically on our ability to infer and to predict, relates to our capacity to 'read' the minds of other people.

Mind-reading

All human interaction is possible only because we understand that we and other people have an executive mental capacity behind the scenes – a mind that cannot be seen, but which directs all our observable actions. Our social dealings require us to try to read the minds of others, to understand and to contextualise behaviour and communication by ascribing belief and motivation to them. Cognitive scientists use the expressions 'mind-reading' and 'Theory of Mind' (ToM) to refer to our ability to attribute states of mind to ourselves and to others, and thereby to interpret and predict behaviour and to negotiate all social dealings with other people.[25] We use this capacity to understand the behaviour of our fellow humans (discernible also in members of other species: the phrase 'theory of mind' was originally devised to explain the mind-reading abilities of chimpanzees[26]). Even though we cannot know for sure what is going on in someone else's mind, our Theory of Mind allows us to assume, from observable behaviour, that 'other individuals want, think, believe'.[27]

23 Sutton, 'Exograms and Interdisciplinarity', p. 212.

24 Distributed cognition is widely studied in Philosophy and Psychology: see Daniel C. Dennett, *Consciousness Explained* (Harmondsworth, 1991); James V. Wertsch, *Voices of the Mind: A Sociocultural Approach to Mediated Action* (Cambridge, MA, 1991); Pascal Boyer and James V. Wertsch, *Memory in Mind and Culture* (Cambridge, 2009), especially Chapter 5, 'Collective Memory'; and Clark, *Supersizing the Mind*.

25 A good brief overview of ongoing discussions of everyday theorising about the mind, with a substantial bibliography, is by the developmental psychologist, Alison Gopnik, 'Theory of Mind', in *MIT Encyclopaedia of the Cognitive Sciences*, ed. Robert A. Wilson and Frank C. Keil (Cambridge, MA and London, 1999), pp. 838–40. For a recent account of the cognitive basis of mind-reading that posits a two-tier combination of high- and low-level processes, see Ian Apperly, *Mindreaders: The Cognitive Basis of 'Theory of Mind'* (Hove and New York, 2011). For more extensive and specialised discussions, see Simon Baren-Cohen, *The Maladapted Mind: Classic Readings in Evolutionary Psychology* (Hove, 1997); Shaun Nichols and Stephen Stich, *Mind Reading: An Integrated Account of Pretence, Self-Awareness, and Understanding Other Minds* (Oxford, 2003); and Bertram F. Malle, *How the Mind Explains Behaviour: Folk Explanations, Meaning, and Social Interaction* (Cambridge, MA, and London, 2004).

26 David Premack and Guy Woodruff, 'Does the Chimpanzee have a Theory of Mind?', *Behavioural and Brain Sciences* 1 (1978), 515–26.

27 Ibid., p. 525.

This evolved capacity, or social adaptation, allows us to interpret observable behaviour as being motivated by underlying mental process: thoughts, emotions, desires and intentions. For example, if someone rises but does not alight when you step onto a bus, you might assume that a seat is being offered to you, apparently out of courtesy or sympathy for apparent advanced years or incapacity: the mental process is back-filled from the observed behaviour, and thoughts and even perhaps feelings are imagined to make sense of the behaviour. Of course, the assumption could be wrong: the person might have wanted instead to stretch a cramped limb, or to pursue some other plan that is not immediately obvious to an observer, or was not imagined on that occasion. We read the situation and the individual behaviour, taking context and prior instances into account, but the process is a form of deduction that relies on earlier experience in conjunction with observation. Such knowledge-building is cumulative, and inexact: we build up mental models of people and observable behaviour as a result of experience, then use those models in future encounters, not always fully successfully.[28] Given the huge amount of non-verbal communication that occurs, negotiating interactions with other people via the activation of such assumptions and the imagination of mental states in others, is necessary, but can be difficult, and is therefore often incomplete or imperfect. So much guesswork is required, as well a reliance upon our knowledge, selection and applications of behavioural norms, and everyday correlations between action and psychological motivation.

This human cognitive ability is a universal aspect of the evolved mind, so we all use our Theory of Mind all the time, but usually this process occurs below the horizon of conscious awareness or intention. But the human facility for mind-reading is by no means consistently accurate or reliable; it is an approximate skill at best. We read minds wrongly or incompletely, when we misattribute intentions and feelings, or incompletely understand complex psychological processes in our intersubjective relations. When this faculty is impaired, often through autism or Asperger's syndrome, 'mindblindness' results: the limited or absent Theory of Mind, with resultant inadequate social negotiation skills.[29] This spectrum of cognitive impairment results in a reduced or inadequate or absent ability to imagine the thoughts and feelings of other people, a deficiency that has wide-reaching effects on human development and social functioning.

Mind-reading in action brings the culture-cognition inter-reliance into sharp focus. As with all activities that relate to our interactions with other humans, mind-reading is subject to cultural contingency, as anyone who has travelled to a foreign culture or even interacted with an unfamiliar social

[28] This idea is similar to that at the base of Schema Theory, on which, see the foundational work Roger Schank and Robert Abelson, *Scripts, Plans, Goals and Understanding: An Inquiry into Human Knowledge Structures* (Hillsdale, NJ, 1977). For the application of schema theory to literary texts, see Elena Semino, *Language and World Creation in Poems and Other Texts* (London, 1997), and Jonathan Culpeper, *Language and Characterisation* (London, 2001).

[29] Simon Baron-Cohen, *Mindblindness: An Essay on Autism and Theory of Mind* (Cambridge, MA, and London, 1995).

group will have noticed. More precisely, social psychologists have argued that our Theory of Mind is not only influenced by cultural content, but is firmly based in socio-cultural practices, embedded in 'folk psychological narratives' acquired during infant development, as Daniel Hutto claims.[30] He goes on to link this development with storied patterns of behaviour: 'it is through scaffolded encounters with stories of the appropriate kind that children learn how the core propositional attitudes behave with respect to one another and other standard mental patterns'.[31]

Similarly, neuropsychologists have quite recently begun to explore the interaction of culture and cognition, specifically how cultural forces have an impact on mental functioning in the emerging field known as Cultural Neuroscience.[32] For instance, Joshua Goh and Denise Park argue that brain-imaging data provides evidence for 'the joint effects of biological and experiential factors in determining neural development and function over the lifespan'.[33] Similarly, psychologists have demonstrated the impact of culture on the functioning of long-term memory[34] and Theory of Mind.[35]

One of the most interesting areas of research within this field of social cognition treats 'mirror neurons':[36] neurological evidence for the occurrence of mental simulation in response to observed actions. Keyseers and Gazzola summarise the key claims of this line of inquiry: 'While we witness the actions, emotions, and sensations of other individuals, our brain spontaneously recreates a pattern of neural activity that resembles that while we perform similar actions or have similar sensations or emotions.'[37]

Research has shown that primates have the capacity to simulate the cognitive processes predicted from observed behaviour via mirror neurons that 'mediate the mapping of one dimension or representation in a brain onto a completely different dimension'.[38] We not only activate the same neural

30 Daniel Hutto, *Folk Psychological Narratives: The Sociocultural Basis of Understanding Reasons* (Cambridge, MA, and London, 2008), p. x.

31 Ibid. p. xii.

32 The impact is considered to be bi-directional. See the articles in Joan Y. Chiao, ed., *Cultural Neuroscience: Cultural Influences on Brain Function* (Amsterdam, New York, Oxford 2009). For the 'cognitive semantic' study of linguistic evidence for cognitive structures, or an approach to explaining the mind as opposed to the brain, or 'how the functional domain is instantiated in the neural domain', see Ray Jackendoff, *Language, Consciousness and Culture: Essays on Mental Structure* (Cambridge, MA, 2007), p. 3.

33 Joshua Goh and Denise Park, 'Culture Sculpts the Perceptual Brain', in *Cultural Neuroscience*, ed. Chiao, pp. 95–111, at p. 109.

34 Angela H. Gutchess and Allie Indeck, 'Cultural Influences on Memory', in *Cultural Neuroscience*, ed. Chiao, pp. 137–50.

35 C. K. Frank and Elise Temple, 'Cultural Effects on the Neural Basis of Theory of Mind', in *Cultural Neuroscience*, ed. Chiao, pp. 213–23.

36 For the general reader: Marco Iacoboni, *Mirroring People: The New Science of How We Connect with People* (New York, 2008); for the Cognitive Neuroscience perspective: Giacomo Rizzolatti and Corrado Sinigaglia, *Mirrors in the Brain – How Our Minds Share Actions and Emotions*, trans. Frances Anderson (Oxford, 2008); and Jaime A. Pineda, ed., *Mirror Neuron Systems: The Role of Mirroring Processes in Social Cognition* (New York, 2009).

37 Christian Keysers and Valeria Gazzola, 'Unifying Social Cognition', in *Mirror Neuron Systems*, ed. Pineda, pp. 3–37, at p. 32.

38 Lindsay M. Oberman and V. S. Ramachandran, 'Reflections on the Mirror Neuron System: Their Evolutionary Functions beyond Motor Representation', in *Mirror Neuron Systems*, ed.

systems involved in mechanical experience when we observe those behaviours enacted by others, but also 'activate representations of our own emotional states and emotional facial expressions while witnessing the emotions of other individuals'.[39] This neural functioning has been found to play a key role in cultural learning, partly a result of the ability to attribute mental states to other individuals.[40] The resulting ability for the simulation or imitation of inner states, including emotional reactions, allows social interaction to be possible, by means of mind-reading, both in lived experiences, and also in our consumption of representations of human experience in literary encounters.[41]

Theory of Mind is also studied as a social adaptation. Most notably, the evolutionary anthropologist Robin Dunbar[42] and his collaborators argue that the superior brain size and capacity of primates is a biological adaptation to meet the functional demands of 'bonded social systems'.[43] The implication is that our evolved brain capacity allows us to hold in our minds several levels of intentional states–embedded wishes, perceptions, beliefs and mental states–as functional necessities of interacting in complex social networks.[44] More broadly, evolutionary anthropologist Michael Tomasello provides the 'cultural intelligence hypothesis' to explain why human societies are far more complex that those of other species: 'human beings collaborate with, communicate with, and learn from their groupmates based on unique skills and motivations for shared intentionality and cultural learning'.[45] Scholars have developed arguments along these lines for distributed cognition and the fundamentally intersubjective nature of human experience, including as it is made possible through shared fictional narratives.[46] Similarly, the psychologist Michael Corballis argues that the distinguishing feature of the human mind is its capacity to embed thoughts within other thoughts – recursion – which allows for the creation and communication of complex ideas, narratives and social relationships.[47]

Pineda, pp. 39–59, at p. 41.

39 Keysers and Gazzola, 'Unifying Social Cognition', p. 4.

40 E. A. R. Losin, M. Dapretto and M Iacoboni, 'Culture in the Mind's Mirror: How Anthropology and Neuroscience can Inform a Model of the Neural Substrate from Cultural Imitative Learning', in *Cultural Neuroscience*, ed. Chiao, pp. 175–90, at p.186.

41 A phenomenon acknowledged in Rizzolatti and Sinigaglia, *Mirrors in the Brain*: 'this sharing [of ideas and emotion via 'bodily sounds and movements'] is the basis on which the theatre evolves and revolves, and mirror neurons, which become active both when an individual executes and act and when he observes it being executed by others, now provide this sharing with a biological explanation' (p. ix).

42 Robin Dunbar, 'Brain and Behaviour in Primate Evolution', in *Mind the Gap: Tracing the Origins of Human Universals*, ed. Peter M. Kappeler and Joan B. Silk (Heidelberg, Dordrecht, London, New York, 2010), pp. 315–30; and Dunbar, 'The Social Brain Hypothesis', *Evolutionary Anthropology* 6 (1998), 178–90.

43 Dunbar, 'Brain and Behaviour', p. 315.

44 Robin Dunbar, 'The Social Brain: Mind, Language, and Society in Evolutionary Perspective', *Annual Review of Anthropology* 32 (2003), 163–81.

45 Michael Tomasello, 'The Gap is Social: Human Shared Intentionality and Culture', in *Mind the Gap*, ed. Kappeler and Silk, pp. 331–50, at p. 346. See also Tomasello, *The Cultural Origins of Human Cognition* (Cambridge, MA, 1999), 'Understanding and Sharing Intentions: The Origins of Cultural Cognition', *Behavioral and Brain Sciences* 28.5 (2005), 675–91.

46 Jordan Zlatev, T. P. Racine, C. Sinha and E. Itkonen, eds, *The Shared Mind: Perspectives on Intersubjectivity* (Amsterdam, 2008).

47 Michael C. Corballis, *The Recursive Mind: The Origins of Human Language, Thought and*

Theory of mind and literary studies

As these studies from a range of different fields show, mind-reading is now a well-known concept in Cognitive Science, though, notwithstanding its application to represented behaviour, it has come to be used in relation to literary texts only recently, where the added complication of the narrator, and therefore the representation or reporting of mind-reading, introduces a further dimension of complexity to the issue. But the powerful capacity of literature to represent consciousness, and the internal dimension of intersubjective experience, has a lot to offer the discussions on Theory of Mind, both in terms of its capacity to represent mind-reading, and in its requirement for that capacity in its use. On the one hand, literature can show human observations, and it can represent the minds of others, and the mind-reading that goes on in those minds. On the other, Theory of Mind, a function of the adapted capacity for social cognition, is necessary to the understanding of literature, in perceiving cause and effect in terms of motivated human behaviour.[48]

Using ideas and methods from Cognitive Science, scholars working in Cognitive Cultural Studies have developed new ways of identifying and analysing the representation of mind-reading in literary texts, and the use of that facility in the consumption of literature. Literature, Zunshine argues, feeds our 'hungry' Theory of Mind,[49] allowing us to exercise our mind-reading skill, and is in turn essential for the existence of literature because it allows us as readers to fill out into fully human-like beings the characters briefly sketched in words: 'The very process of making sense of what we read appears to be grounded in our ability to invest the verbal flimsy constructions that we generously call 'characters' with a potential for a variety of thoughts, feelings, and desires and then to look for the 'cues' that would allow us to guess at their feelings and thus predict their actions.'[50] Similarly, Natalie Phillips shows how readers use their theory of mind during the process of interpretation, especially in relation to apprehending distinctive habits of mind as a component of characterisation and apportioning interest and empathy.[51]

This process requires the investment of cognitive effort and also emotional investment on the part of the reader, a process that 'mirror neuron' research can tell us actually involves simulation, and in turn can have a developmental effect. As Keith Oatley expresses this process, 'we project ourselves into the

Civilization (Princeton, 2011).

48 Robin Dunbar, *Grooming, Gossip, and the Evolution of Language* (Cambridge, MA, 1999), p. 102.

49 Lisa Zunshine, 'Lying Bodies of the Enlightenment: Theory of Mind and Cultural Historicism', in *Introduction*, ed. Zunshine, pp. 115–33, at p. 120.

50 Lisa Zunshine, *Why We Read Fiction: Theory of Mind and the Novel* (Columbus, 2006), p. 10; also in Zunshine, 'Theory of Mind and Experimental Representations of Fictional Consciousness', *Narrative* 11 (2003), 270–91, at p. 274.

51 Natalie Phillips, 'Distraction as Liveliness of Mind: Cognitive Approach to Characterization in Jane Austen', in *Theory of Mind and Literature*, ed. Paula Leverage, Howard Mancing, Richard Schweickert and Jennifer Marston William (West Lafayette, IN, 2011), pp. 105–22.

simulated world and we may come back changed'.[52] Oatley's work on the development of empathy and emotional reactions more broadly has made a significant contribution to our understanding of the engagement of emotions during literary encounters, and how little those responses differ from those creates in real-life situations.[53] Like other Cognitive Cultural approaches, Oatley's has shown how fruitful a robust interaction between the sciences and literary analysis can be, especially in understanding the development of the novel,[54] and the way in which literature can be both an instrument of cognition and also a tool for refining mind-reading abilities.[55]

Some of these ideas sound like they are similar to those held by Literary Darwinists, who claim an important role for literature in human evolution, but most Cognitive Cultural theorists do not subscribe to such a view. Literary Darwinists argue that the social and psychological functions of art can be explained from an evolutionary perspective and that literature can be traced as an adaptation.[56] Scholars working within Cognitive Cultural Studies dispute the validity of such nativist claims that any so-called literary disposition – an evolved adaptation or innate competence for creating and consuming literary texts – can be seen to have played a role in natural selection.[57] Instead, they propose to explore how cognitive functioning has an effect on culture, and concentrate on how minds can be shown to function in specific cultural contexts.

Scholars working within the frameworks developed by Cognitive Cultural Studies have suggested that humans use cognitively evolved abilities to make sense of fictional narrative, as well as real-life encounters, through deploying Theory of Mind, and that literary encounters serve the function of fine-tuning this technique of attributing mental states to other people as well as providing the pleasure of '"trying on" mental states *potentially*

52 Keith Oatley, 'Theory of Mind and Theory of Minds in Literature', in *Theory of Mind and Literature*, ed. Leverage *et al.*, pp. 13–26, at p. 23.

53 See also Patrick Colm Hogan, 'On Being Moved: Cognition and Emotion in Literature and Film', in *An Introduction*, ed. Zunshine, pp. 237–56.

54 See also Zunshine, 'Fiction and Theory of Mind: An Exchange', *Philosophy and Literature* 31.1 (2007), 189–96; 'Theory of Mind and Fiction of Embodied Transparency', *Narrative* 16.1 (2008), 65–92; and 'Lying Bodies of the Enlightenment'.

55 See Oatley, 'Theory of Mind', pp. 13–26. The latter idea – that literary texts can assist in the refinement of cognitive tools for making sense of the world beyond the text – is developed in a similar way in relation to *Beowulf* by one of the few examples of the analysis of Old English textual narrative structures as evidence of narrative thinking: David Herman and Becky Childs, 'Narrative and Cognition in *Beowulf*', *Style* 37 (2003), 177–203.

56 See the foundational work of Joseph Carroll, *Literary Darwinism: Evolution, Human Nature and Literature* (London and New York, 2004), and also Jonathan Gottschall and David Sloan Wilson, eds, *The Literary Animal* (Evanston, IL, 2005); Dennis Dutton, *The Art Instinct: Beauty, Pleasure and Human Evolution* (Oxford and New York, 2009); and Brian Boyd, *On the Origin of Stories: Evolution, Cognition and Fiction* (Cambridge, MA, and London, 2009). On the evolutional benefits of fiction from this perspective, see also Paul Hernadi, 'Why is Literature? A Coevolutinary Perspective on Imaginative Worldmaking', *Poetics Today* 32.1 (2002), 21–42; Brian Boyd, 'Fiction and Theory of Mind', *Philosophy and Literature* 30 (2006), 571–81.

57 For a detailed critique along these lines, see J. Kramnick, 'Against Literary Darwinism', *Critical Inquiry* 37.2 (2011), 315–47.

available to us'.[58] Indeed, one of the chief sources of enjoyment in any literary encounter might be what Alan Palmer describes as 'being told what a variety of fictional people are thinking',[59] a more detailed account of human cognitive and emotional experience than we have in our everyday interactions with living people. In this view, we read fiction because it 'engages, teases, and pushes to its tentative limits our mind-reading capacity', precisely because it requires us to keep track of the complex levels of intentionality, or layers of mental states, and sources of representations.[60]

These arguments recognise the cultural embeddedness of mind-reading and its representation in literature, but not the impact of significant diachronic change. The majority of investigations into 'how we construct our world both culturally *and* cognitively'[61] focus on exploitations of our mind-reading ability in the novel. Apart from Zunshine, scholars, such as Alan Richardson,[62] Ellen Spolsky,[63] and Blakey Vermeule[64] have studied Theory of Mind in early-modern and contemporary literature, but relatively little work has been done on medieval texts using cognitive approaches. The focus on the novel is entirely understandable, as the late eighteenth century was undoubtedly the time in the English tradition at which complex levels of intentionality were ascribed to fictional characters, of the sort that 'I know that you know that I know',[65] a rhetorical structure used in the title of George Butte's well known study, *I Know that You Know that I Know: Narrating Subjects from Moll Flanders to Marnie,* in which he studies 'the way human subjects narrate and get themselves narrated'.[66] Butte claims to have identified a precise historical moment when a change in narrative representations of consciousness occurred – the development of what he calls 'deep intersubjectivity' (or 'partially interpenetrating consciousnesses') in the early nineteenth century.[67]

This idea is very similar to that of Theory of Mind, though Butte does not use the latter phrase. He does make the claim, though, that this literary facility makes its first appearance around the time of, and perhaps specifically in the work of, Jane Austen. Lisa Zunshine seems to be of a similar opinion, though she does make the point that Butte's 'deep intersubjectivity' is what cognitive scientist knows as Theory of Mind. In her article 'Why Jane Austen

58 Zunshine, *Why We Read Fiction*, p. 17.
59 Alan Palmer, *Fictional Minds* (Lincoln and London, 2004), p. 10.
60 Zunshine, *Why We Read Fiction*, p. 4.
61 Lisa Zunshine, 'Richardson's *Clarissa* and a Theory of Mind', in *The Work of Fiction: Cognition, Culture and Complexity*, ed. Alan Richardson and Ellen Spolsky (Aldershot and Burlington, 2004), pp. 127–46, at p. 142.
62 Alan Richardson, 'Studies in Literature and Cognition: A Field Map', in *The Work of Fiction*, ed. Richardson and Spolsky, pp. 1–30.
63 Ellen Spolsky, 'Women's Work is Chastity: Lucretia, *Cymbeline*, and Cognitive Impenetrability', in *The Work of Fiction*, ed. Richardson and Spolsky, pp. 51–84.
64 Blakey Vermeule, *Why Do We Care about Literary Characters?* (Baltimore, 2010).
65 Lisa Zunshine, 'Why Jane Austen was Different and Why We May Need Cognitive Science to See it', *Style* 41.3 (2007), 275–99, at p. 279, citing Butte.
66 George Butte, *I Know that You Know that I Know: Narrating Subjects from Moll Flanders to Marnie* (Columbus, 2004), at p. vii.
67 Ibid., p. 27.

was different and why we may need Cognitive Science to see it', Zunshine takes the literary use of Theory of Mind back to the eighteenth century, and ties it even more closely to the rise of the novel. Zunshine and Butte both dwell on a particular instance in *Persuasion*, as typical of Austen's reliance on mind-reading:

> It did not surprise, but it grieved Anne to observe that Elizabeth would not *know* [Wentworth]. She *saw* that he *saw* Elizabeth, that Elizabeth *saw* him, that there was complete internal *recognition* on each side; she was *convinced* that he was ready to be *acknowledged* to be an acquaintance, *expecting* it, and she had the pain of *seeing* her sister turn away with unalterable coldness.[68]

Jane Austen is certainly one of the most psychological of novelists (and probably one of the earliest to reflect on this subject). We can see in this extract multiple levels of embeddedness. In view of such evidence, Zunshine agrees with Butte's idea that in Austen, we see 'a new representation of those subjectivities as layered in a specific way',[69] and claims the deep compatibility of Butte's argument with her own view, coming from within the field of Cognitive Cultural Studies. She says that 'his argument provides a crucial first step for recognising Austen's prose as actively experimenting with readers' cognition'.[70] Similarly, Kay Young, in her recent study of Austen, Eliot and Hardy, *Imagining Minds*, argues that 'the novel is an aesthetic map to and experience of the nature of the mind-brain'.[71]

There is undoubtedly something decidedly different and new about Jane Austen's narrative method, and her engagement with emerging brain-science, particularly in a consideration of the relationship between the mind and the body.[72] It is abundantly clear that the deployment of mind-reading in nineteenth-century literature is not in dispute, though its innovative quality is open to question. So little work has been done on literature other than early-modern texts that it is difficult to make informed comments about earlier literary uses of Theory of Mind. Indeed, in *Why We Read Fiction*, Zunshine even goes so far as to claim that only novelists work a reader's mind-reading ability to maximum capacity, and, further, that it would have been impossible to represent a Theory of Mind as complicated as we find in eighteenth-century fiction in a medieval text such as the Old English poem *Beowulf*, even though all texts require us to use our Theory of Mind.[73] She claims that medieval texts do not experiment with metarepresentation (the representation of representation) and Theory of Mind to the same extent that more recent ones do, constrained by the 'material realities

68 Jane Austen, *Northanger Abbey and Persuasion*, ed. R. W. Chapman (London, 1969), p. 117, cited in Zunshine, 'Why Jane Austen was Different', p. 275.

69 Zunshine, 'Why Jane Austen was Different', p. 276, and Butte, *I Know that You Know*, p. 4.

70 Zunshine, 'Why Jane Austen was Different', p. 276.

71 Kay Young, *Imagining Minds: The Neuro-Aesthetic of Austen, Eliot, and Hardy* (Colombus, 2010) p. 9.

72 See Antonina Harbus, 'Reading Embodied Consciousness in *Emma*', *Studies in English Literature 1500–1900* 51.4 (2011), 765–82.

73 Zunshine, *Why We Read Fiction*, pp. 73–5.

of its time's textual reproduction'.[74] Presumably, by this statement, Zunshine implies that texts created for oral performance rather than via mass print production are limited in the number of nested mental states that can be communicated, constrained by the more limited capacity of auditory memory in that they 'may never be able to embed more than three layers of intentionality',[75] compared with five or six in texts such as *Mrs Dalloway*. To demonstrate this claim, Zunshine cites the instance in that novel where Mrs Bruton asks for Richard's and Hugh's assistance to write to the editor of the *Times*, where she lists five levels of intentionality: 'Richard *suspects* that Lady Bruton indeed *believes* that because, as Hugh *says*, the makers of the pen *think* that it will never wear out, the editor of the Times will *respect* and publish the ideas recorded by his pen.'[76] This citation shows how hard the reader has to work to keep track of the line of nested ideas in this scene, and while it is likely that Woolf relied on such complicated inferential systems in her prose, the presence or absence of such systems in earlier literature is assumed by Zunshine rather than explored.

Notwithstanding the perceived constraints in the text, Zunshine finds some evidence of 'experimentation with our metarepresentational ability' – or ability to keep track of who thought and said what – in *Beowulf*, specifically the unpredicatability of the reader's interpretation of the characters' behaviour and motivations in Unferth's apparent change of heart and presentation of the sword to Beowulf. She says: 'Our interpretation of Unferth's and Beowulf's behavior and personality *will certainly be structured* by our metarepresentational ability: for the poem calculatedly feeds this ability by implying, first, that there is an important difference between the states of Unferth's mind *then* and *now*, and, second, that Unferth's opinion matters to a certain degree within the social world of the poem'.[77]

While there is nothing to dispute in this statement, it relies on a light and not entirely accurate reading of the text. For instance, the claim that 'Unferth, who (we infer, using our ToM) must be jealous of the attention and respect heaped on the newcomer'[78] ignores other information provided in the text and indeed encoded in Unferth's name, on which there is a great deal of critical discussion.[79] Similarly, the ritualistic or heavily symbolic systems of meaning embedded in this text (such as Beowulf's sword), like many other medieval ones, require more deductive work and different practices of interpretation than encounters with modern novels do.

[74] Ibid., p. 73.
[75] Ibid.
[76] Ibid., p. 33.
[77] Ibid., p. 74.
[78] Ibid., p. 73.
[79] For example, M. F. Vaughan, 'A Reconsideration of "Unferth"', *Neuphilologische Mitteilungen* 77 (1976), 32–48; Carol Clover, 'The Germanic Context of the Unferth Episode', *Speculum* 55 (1980), 444–68; Jane Roberts, 'Old English *un-* "very" and Unferth', *English Studies* 61 (1980), 289–92; R. D. Fulk, 'Unferth and his Name', *Modern Philology* 85 (1987), 113–27; and Fred C. Robinson, 'Personal Names in Medieval Narrative and the Name of Unferth in Beowulf', in Robinson's *The Tomb of Beowulf and Other Essays on Old English* (Oxford, 1993), pp. 218–23.

Indeed, a great deal of scholarship on the so-called digressions in *Beowulf* and its synthetic, multi-layered communication of ideas – 'its kaleidoscopic effect',[80] would suggest that the recipient of this text is called upon to hold several levels of intentionality in mind during textual processing. So, certainly, a medieval text might experiment with our metarepresentational ability differently from the way in which modern novels do, and use distinctive systems of embedded intentionalities, but the differences have not yet been studied and are far from having been demonstrated.

Other scholars focusing on modern texts tend to agree with Zunshine about the special capacity of the novel for making cognitive demands on readers, and take her assumptions as settled. Howard Mancing is an exception to this trend; indeed he vigorously challenges the notion that advanced skills arising from Theory of Mind were evidently required for pre-modern textual encounters. Mancing makes the argument that *Don Quixote*, which he argues is an early example of the novel, demonstrates the need for the 'multi-leveled presentation of fictional minds'.[81] Similarly, Paula Leverage has argued that Chrétien de Troyes demonstrates an interest in Theory of Mind, but she is one of very few scholars working with Theory of Mind and medieval texts.[82]

I would take this matter further, to consider the implications for textual understanding and the future of cognitive studies of literature and culture more broadly. The restriction to early-modern and modern texts overlooks the potential of this approach for bringing historical depth and cultural variability to the study of the Theory of Mind and levels of intentionality. There remains, then, an opportunity to broaden the diachronic scope of the inquiry into literary interest in Theory of Mind to the earlier medieval period, specifically, to explore the applicability and potential insights available from these approaches to Old English poetry. These, the earliest English texts, can be shown to demonstrate not only an interest in Theory of Mind, but also a narrative technique that relies on the representation of multiple embedded consciousnesses and levels of intentionality.

As the first exemplary case study of this key idea from Cognitive Cultural Studies in practice, the following discussion will analyse the Theory of Mind implied by the narrator of *The Dream of the Rood*, a text that relies on several complex mental processes: a sophisticated interpretive response involving conceptual blending and multiple levels of intentionality; familiarity with the

80 See Robert E. Bjork, 'Digressions and Episodes', in *A Beowulf Handbook*, ed. Robert E. Bjork and John D. Niles (Lindoln, NE, 1997), pp. 193–212, at p. 211. Bjork, in fact, likens the effect to that produced by Virginia Woolf. See also *Klaeber's Beowulf*, 4th edn, ed. R. D. Fulk, Robert E. Bjork and John D. Niles (Toronto, 2008), 'Non-Linear Narration', pp. xcv–xcix.

81 Howard Mancing, 'Sancho Panza's Theory of Mind', in *Theory of Mind*, ed. Leverage *et al.*, pp. 123–32, at p. 127. Mancing explains: 'My point throughout this essay is that a profound understanding of human consciousness and Theory of Mind has been a characteristic of literature for more than four centuries' (p. 130).

82 Paula Leverage, 'Is Perceval Autistic? Theory of Mind in the *Conte del Graal*', in *Theory of Mind*, ed. Leverage *et al.*, pp. 133–51. See also Leverage, *Memory and Reception: A Cognitive Approach to the Chansons de Geste* (Amsterdam, 2010).

representation of images within a dream context; the decoding of symbolism and intertextual references; and possession of a mind schema responsive to the heuristic function of poetry. This text has been selected in order to provide a contrastive study of a different approach presented in chapter 4 above, but more specifically because as a dream vision, its highly visual qualities and narrative complexity featuring multiple embedded narrators makes it an ideal focus for a study of mind-reading.

This poem can be analysed to demonstrate mind-reading in action via the figure of the dreamer-narrator, and also to prompt the same activity in the mind of the recipient in response to the text itself. Other examples of mind-reading in action in Old English poetry are drawn from two other texts, *Elene* and *Beowulf*, to demonstrate that the dream vision is not an isolated example, but rather that Anglo-Saxon writers routinely represented Theory of Mind in action, deployed multiply embedded mental states, and called upon their audiences to use, during textual reception, advanced mental processing similar to that required of later audiences. This evidence can show the cultural process of cognition at work over a much more extensive timescale than is usually considered, and allow us to start exploring at very long range the negotiated border between stable mental processes and variable aspects of culture.

Mind-reading in The Dream of the Rood

The Dream of the Rood depicts mind-reading, metarepresentation and multiple levels of intentionality in a highly complex manner that challenges any suggestion that these qualities characterise the modern novel in distinctive ways. This imaginative, didactic dream vision features a talking, chameleon Christian Cross that recounts the Crucifixion from its own point of view. This Cross is endowed in this text with not only its usual Christian symbolism, but also with a human consciousness likened to that of Christ; it also bears similarities to that of the dreamer-narrator. This consciousness is the key link between the Cross and Christ, but also between it and the dreamer and ultimately the recipient of the text specifically through common mind-reading abilities. All three textual entities – Cross, Christ and dreamer – interact, and indeed overlap, primarily through their demonstrated mind-reading abilities, a demonstrated mental process that is modelled for the reader or hearer of the text. The poem invites the recipient to use mind-reading skills to observe and process the conceptual content of the narrative account of the Crucifixion, and also to process its didactic message. The rhetorical power of the text can be seen to rely on the specifically conceptual processing and indeed reorganisation that its interpretation requires.

This interpretive approach is invited through customised narrative arrangement, motivated to produce a cognitively synthetic engagement with the text triggered by the new perspective and the shifting images presented. As Éamonn Ó Carragáin has shown so fully, the narrative of *The Dream of the Rood* is thoroughly innovative: it departs from all four Gospels; is the only medieval text to have recounted the Crucifixion from the Cross's perspective; and the only one to have focused on the awful dilemma of the Cross called upon by its lord to become an instrument of that same lord's torture and death.[83] Ó Carragáin claims that this unique text therefore 'adapts the narrative to the devotional needs of a private reader',[84] an argument that can be advanced by scrutinising more closely, with the tools provided by Cognitive Cultural Studies, the conceptual demands it makes on its reader.

The reader or hearer of even an emphatically visual Old English poem, such as *The Dream of the Rood*, is certainly required to create images and systems of representations in the mind, to blend concepts, and to entertain a belief in revelatory dreams, a talking cross and a Christian vision of salvation. More specifically, the audience must navigate multiple embedded consciousnesses – to keep track of reports of observations, and observations of observations, as well as literary self-reference. To make sense of the text, the reader or hearer audience must witness and engage in mind-reading: to predict states of mind from behaviour, and from representations of behaviour.

On the most obvious level are the many reported predictions of mental states by the Cross and the dreamer. Near the opening it is narrated that the dreamer saw angels observing the tree and interpreting it as an object that is 'not a gallows' – by implication the exact opposite (9b–12):

Beheoldon þær engel Dryhtnes ealle,
fægere þurh forðgesceaft. Ne wæs ðær huru fracodes gealga,
ac hine þær beheoldon halige gastas,
men ofer moldan ond eall þeos mære gesceaft.

All those beautiful by eternal decree beheld there the angel of the Lord – indeed that was not there the gallows of a wicked one – but the holy souls beheld it there, men over the earth, and all this glorious creation.

The slight awkwardness of these lines has attracted much commentary and emendation[85] (this is Krapp and Dobbie's punctuation),[86] with some critics arguing that 'engel Dryhtnes' refers to Christ[87] and others that an emendation (usually to 'englas') is essential for sense and theological orthodoxy.[88] But part of the problem lies in the dual focalisation of the 'not gallows' object, the Cross, interpreted as something decidedly other than an instrument of torture by the dreamer and all creation. Putting aside the difficulty of 'engel',

83 Éamonn Ó Carragáin, *Ritual and the Rood: Liturgical Images and the Old English Poems of the 'Dream of the Rood' Tradition*, British Library Studies in Medieval Culture (London, 2005), pp. 3–7.

84 Ibid., p. 9.

85 Murray McGillivray, '*Dream of the Rood* 9–12 and the Christian Liturgy', *Notes and Queries*, n.s. 52 (2005), 1–2; and Ó Carragáin, *Ritual and the Rood*, p. 325.

86 Bruce Mitchell, '*Dream of the Rood* Repunctuated', in *Words and Works: Studies in Medieval English Literature in Honor of Fred C. Robinson*, ed. Nicholas Howe and Peter S. Baker (Toronto, 1998), pp. 143–57.

87 See Jeremy I. Wheelock, 'The Word Made Flesh: 'Engel Dryhtnes' in *The Dream of the Rood*', *English Language Notes* 37.3 (2000), 11, and Ó Carragáin, *Ritual and the Rood*, p. 325.

88 Paul Cavill, '"Engel Dryhtnes" in "The Dream of the Rood" 9b Again', *Neuphilologus Mitteilungen* 86 (1985), 287–92.

the thought that the Cross is something other than a gallows goes beyond the dreamer's own personal interpretation. Here the dreamer is predicting the states of mind of all creation, and equating that interpretation with his own: he is using his Theory of Mind to explain observed behaviour, and thereby embedding several levels of intentionality within a particularly didactic moment in the text. The logic and embedded mental states run as follows: the reader *knows* that the narrator *suggests* that the dreamer *understands* that the angels and all creation *interpreted* the cross as something other than a gallows (and the reader is invited to *interpret* the image likewise). Reader mental alignment is encouraged through the display of the cognitive functioning of the dreamer to reach this point of conceptual nesting.

This is just one of many instances where the didactic impulse shows itself through modelling to the reader an idealised cognitive act – here, mental representation and the interpretation of symbols. Throughout the opening section of the poem, this process recurs. For instance, the Cross's external appearance, as it is reported by the dreamer, embodies the perceived interpretations of observers, who see the Cross as symbolising either torture and death as a gallows, or salvation, as a Christian Cross: 'geseah ic þæt fuse beacen / wendan wædum ond bleom' (21b–2a), 'I saw that eager beacon change its coverings and colours'. The wording of this line attributes intentionality and emotion to the Cross, whose chameleon appearance is motivated and controlled, and interacts with its capacity to act as if it were human. Indeed, the reader is invited to interpret it simultaneously as both human-like (with recognisable feelings, behaviours and cognitive functioning) and also as a rich site of symbolic meaning, a complex and multi-layered act of reading that holds apparently incompatible ideas in tandem and relies extensively on the reader's acceptance of the Theory of Mind attributed by the dreamer to the Cross as a dominant factor in the creation of coherence.

That coherence relies not just on the representation of the dreamer's acts of mind-reading on the Cross and on those observing the Cross, but also on representations of the Cross itself in the act of mind-reading. The Cross attributes mental states to the humans who initially felled it, making inferences about motivation from observable and received actions (30b–31):

Genaman me þær strange feondas,
geworhton him þær to wæfersyne,   heton me heora wergas hebban.

Powerful enemies seized me there, fashioned me there as a spectacle for themselves, required me to hold up their criminals.

These lines silently presuppose and conflate the motivations of the woodcutter, carpenter and executioner, in an act of mind-reading that singles out an intention to coerce an innocent and unwilling tree to become an instrument of death, an associate of criminals and a form of entertainment. This attributed intention performs the rhetorical function of implying that the tree has been wronged, not only through action, but because its feelings and sensibilities were not consulted and respected – mind-reading was absent by the woodcutter, carpenter and executioner. The reader constructs the idea of this omission in the context of having performed some mind-reading of their own in the act of textual interpretation, by analogy with the dreamer's mind-reading of the Cross and others. As a result of supplying to it a mind that informs actions, the reader has endowed the Cross with the acute powers of emotion, thoughts and intentions necessary to make sense of the Cross's narrative of the Crucifixion. The pattern of mind-reading that results in the mind of the reader is of the kind: I *know* that the dreamer *knows* that the Cross *knows* that the woodcutter, carpenter and dreamer *did not know*.

Similarly, the Cross attributes emotions and intentions to Christ, in an act of mind-reading that supports the alignment of the two created throughout the text. He backfills the intention that motivates haste: 'geseah ic þa frean mancynnes / efstan elne micle, þæt he me wolde on gestigan' (33b–4), 'I saw then the Lord of mankind hasten with courage, in that he intended to ascend me'. Later too, he refers to Christ as 'stiðmod' (40a), 'unflinching', in a compound term containing 'mod' ('mind', but also 'heart, spirit, and even thought and mood'), a linguistic characteristic of Old English adjectives that foregrounds the mind in emotional states.[89] Elsewhere, too, the Cross attributes behaviour to feelings. After the Crucifixion, Christ rested 'meðe æfter ðam miclan gewinne' (65a), 'tired after the great struggle'. The Cross also attributes beliefs and intentions to God. In likening itself, chosen over all trees, to Mary, chosen over all women (90–4), it claims that 'me þa geweorðode wuldres ealdor' (90), 'the guardian of the world honoured me then'. The reader supplies the multiple levels of mental processing: I *know* that you *know* that He *knows*.

The Cross goes further in this chain of knowing when it defines its symbolic power in the minds of believers: 'ond ic hælan mæg / æghwylcne anra, þara þe him bið egesa to me' (85b–6), 'and I may save each of those in whom there is fear of me'. It even predicts the future mental states of people at the Day of Judgement (115–16):

Ac hie þonne forhtiað, ond fea þencaþ
hwæt hie to Criste cweðan onginnen.

But they will fear then and little think what they will begin to say to Christ.

The Cross is not just anticipating behaviour or emotions here, but also intentions and mental states, using his Theory of Mind to put himself into the position of the person on Judgement Day, and to represent to the reader the similar act that this didactic text is crafted to bring about. The point is reiterated when the Cross equates emotions, intentions and salvation (117–21):

Ne þearf ðær þonne ænig anforht wesan
þe him ær in breostum bereð beacna selest,

89 Antonina Harbus, *The Life of the Mind in Old English Poetry* (Amsterdam and New York, 2002), pp. 50–4.

ac ðurh ða rode sceal   rice gesecan
of eorðwege   æghwylc sawl,
seo þe mid wealdende   wunian þenceð.

There is no need then that anyone who already bears the best of beacons in his breast should be afraid, but through the cross each soul from earth that intends to dwell with the ruler will reach the kingdom.

This mind-reading is further modelled by the dreamer at the end of the text when he engages in self-reflexive attribution of a mental state: 'wæs modsefa / afysed on forðwege' (124b–125a), 'my spirit was inspired with longing for the way ahead'. This formulation is characteristic of Old English poetic style, where the mind is explicitly cited as the source of emotions, thoughts and feelings,[90] but also emphatically explicates the link between behaviour and motivated mental processes.

The dreamer-narrator listens to the Cross's account of mind-reading, then does some of his own, not only modelling that behaviour for the reader, but inviting the appropriation of the idealised mental states on display, and indeed necessitating that experimental 'trying on' of a mindset if successful interpretation is to occur, something he does in the poem, in response to the Cross's entreaty (95–7):

Nu ic þe hate,   hæleð min se leofa,
þæt ðu þas gesyhðe   secge mannum,
onwreoh wordum   þæt hit is wuldres beam.

Now I entreat you my dear man that you recount this vision to others, reveal in words that it is the tree of glory.

Its emotional force and didactic impact rely on the creation of readerly empathy. Making sense of *The Dream of the Rood* requires the reader or hearer to grapple with a didactically emphatic message that can be expressed in terms of multiple embedded mental states: I *know* that the author *knows* that the narrator *knows* that the dreamer *knows* that the Cross *knows* that Christ *knows* that I *know*, so it cannot be denied.

Beyond this demonstration of multiple levels of embedded intentionality, this poem and its broader context instantiates the conception of cognition as the 'interface of brain, body, and the social and material world', in Sutton's terms.[91] The existence of part of an earlier version of *The Dream of the Rood* carved in runes upon the surface of the large stone preaching cross, the Ruthwell Cross,[92] not only tells us a great deal about the didactic and social functions of this text, but also indicates contemporary attitudes to the

90 See Harbus, *Life of the Mind*, esp. pp. 31–2.
91 Sutton, 'Exograms and Interdisciplinarity', p. 219.
92 See Ó Carragáin, *Ritual and the Rood*, esp. pp. 58–62; Pamela O'Neill, *'A Pillar Curiously Engraven, with some Inscription upon it': What is the Ruthwell Cross?* British Archaeological Reports, British Series 397 (Oxford, 2005); Calvin B. Kendall, 'From Sign to Vision: The Ruthwell Cross and *The Dream of the Rood*', in *The Place of the Cross in Anglo-Saxon England*, ed. Catherine E. Karkov, Sarah Larratt Keefer and Karen Louise Jolly, Publications of the Manchester Centre for Anglo-Saxon Studies 4 (Woodbridge, 2006), pp. 129–44.

materiality of text and in turn the dynamic interrelationship between poetic concept and the material world. The Cross speaks, both in the poem, and from the surface of this great stone monument. The section of the poem carved into the Ruthwell Cross relates to the courage of Christ in ascending the Cross, and further interacts with the other inscribed words in Latin, and also the images carved onto the cross: of the Crucifixion, but also the Annunciation and the Visitation, as well as other scenes depicting beasts and desert scenes. This interplay between material culture and textual transmission is possible only because the contemporary audiences could accommodate the conceptually complex notion of the Cross as a text to be read – a material and symbolic artefact able to communicate new understanding all the more effectively as a result of representation via blending modes of information modality. In this case, visual and imaginative maps or sources of information need to be cross-mapped and synthesised to make sense of the text, a process that results in the acquisition of new knowledge and enhanced explanatory capacity. As Ellen Spolsky argues in relation to other texts, such 'recategorizaton' among sensory maps when a poem is transfigured into a sculpture facilitates learning and imaginative leaps in our essentially modular minds.[93] The existence of this Cross tells us, then, that complex structures of thought familiar to us prevailed in Anglo-Saxon times, and indeed are presumed in the creation of such artefacts as *The Dream of the Rood* and The Ruthwell Cross and in the conceptual demands they make on their recipients.

The Dream of the Rood, as a vivid dream vision, but also as an inscribed utterance on a preaching cross, embodies the vernacular apprehension and use of complex mental processing. It represents these cognitive activities in action, and calls upon human mind-reading abilities as a textual response. More broadly, this text shows how well developed and indeed sophisticated its audience's understanding and mental processing must have been for such a complex piece of communication to be out in the public realm in different modalities. Other extant texts show that this is not an isolated instance; there is ample evidence to suggest that a certain degree of self-awareness in terms of the mind and its capacities prevailed in Anglo-Saxon England. For instance, use of mind-reading and acceptance of its variability occurs widely elsewhere in the Old English corpus, and occasionally, the subject is treated explicitly. For instance, the hero of *Andreas*, in his attempt to avoid going on a perilous sea journey to rescue Matthew, objects that he does not know the thoughts of the foreigners (198b–201):

Ne synt me winas cuðe,
eorlas elþeodige, ne þær æniges wat
hæleða gehygdo, ne me herestræta
ofer cald wæter cuðe sindon.

The foreign men are not known to me as friends, nor do I know there

[93] Ellen Spolsky, 'Making "Quite Anew": Brain Modularity and Creativity', in *Introduction*, ed. Zunshine, pp. 84–102, at p. 93.

> anything of the thoughts of men, nor are the main paths over the cold water known to me.

This piece of text, apparently not based on anything in the source document,[94] articulates the correlation assumed between habits of thought and cultural familiarity, and in turn, the cultural contingency of adequate mind-reading.

There are many others instances that confirm that mind-reading is not confined to *The Dream of the Rood*. Other, admittedly less striking, instances of Theory of Mind confirm the widespread reliance on this human cognitive capacity and its deep embeddedness in Anglo-Saxon literary culture. Two brief instances, from *Elene* and *Beowulf*, will support this claim.

Mind-reading in Elene

St Helena's shrewd evaluation of Judas's motives and psychological performance is essentially a case of motivated and effective mind-reading. When he is in the process of overcoming a demon, Helena attributes intentions, states of mind, thoughts and wishes to Judas, inferred from his observable behaviour (952a–966):

> Elene gehyrde
> hu se feond ond se freond   geflitu rærdon,
> tireadig ond trag,   on twa halfa,
> synnig ond gesælig.   Sefa wæs þe glædra
> þæs þe heo gehyrde   þone hellesceaþan
> oferswiðedne,   synna bryttan,
> ond þa wundrade   ymb þæs weres snyttro,
> hu he swa geleafful   on swa lytlum fæce
> ond swa uncyðig   æfre wurde,
> gleawnesse þurhgoten.   Gode þancode,
> wuldorcyninge,   þæs hire se willa gelamp
> þurh bearn godes   bega gehwæðres,
> ge æt þære gesyhðe   þæs sigebeames,
> ge ðæs geleafan   þe hio swa leohte oncneow,
> wuldorfæste gife   in þæs weres breostum.

> Helena heard how the fiend and the friend raised strife, the glorious and the evil ones, on both sides, the sinful and the blessed. Her mind was the gladder, because she heard the hell-foe overcome, the ruler of sins, and then marvelled about the wisdom of the man, how he had ever become so full of faith and so understanding, so imbued with perception. She gave thanks to God, to the king of glory, because her will had come about through the son of God in two respects, both in the sight of the tree of glory, and in the faith that she so clearly recognised, the glorious gift in the man's breast.

94 See discussion in Antonina Harbus, 'A Mind for Hagiography: The Psychology of Resolution in *Andreas*', in *Germanic Texts and Latin Models*, Germania Latina IV, ed. K. E. Olsen, A. Harbus and T. Hofstra (Leuven, Paris, Sterling, VA, 2001), pp. 127–40, at p. 131.

Here, we have a cognitive smorgasbord: a description of observed action as perceived through the senses ('heard'), mentally apprehended emotion ('mind was the gladder'), cognitive reaction via imagination, modelling and inference ('marvelled'), cognitive action ('gave thanks'), and the imputation of mental states to Judas ('faith that she clearly recognised . . . in the man's breast'). Helena interprets Judas's behaviour as arising from underlying the complex mental state referred to as 'faith', or belief, and is reported to have read his mind, or imputed mental states to Judas, through an interpretation of his observable behaviour.

Helena's saintly extraordinariness is predominantly shown here through her acute powers of perception and her apparently infallible mind-reading abilities. Any reader of hagiography would recognise the stock account of this figure's exceptional perspicacity and unerring judgment, so the link between excellent mind-reading and divine favour is not new here. Rather, the unexceptional nature of both the hagiographic representation and the Old English expression of it demonstrate how familiar and deeply ingrained mind-reading had become by the Anglo-Saxon period.

Notwithstanding this transmission of rhetorical techniques and representational habits, this text can show, by comparison with its Latin exemplar, particular Old English treatments of mind-reading. The line 'sefa wæs þe glædra' (955b), 'her mind was the gladder' encodes the linguistic predilection for mentioning the mental site of emotional registration as noted above, and perception is foregrounded elsewhere too in a way not found in the original, the *Acta Quiriaci*.[95] Like the *Andreas* example mentioned above, this addition to the source document in its linguistic and cultural translation suggests a special vernacular interest in representing mind-reading in action, and the particular importance placed on that process in human interactions. These two instances suggest that it would be worth exploring whether such inclusions are rhetorically or aesthetically motivated: whether they are part of the unselfconscious attempts by the vernacular translator to communicate the didactic narrative of the original effectively, or reflect a particular local taste for references to mental processing.

Mind-reading in Beowulf

Beowulf provides additional evidence of an Anglo-Saxon capacity for textual representations of complex mind-reading. When Hrothgar tells Beowulf that he has never met anyone whose mind he admired quite so much, he is articulating evaluative mind-reading in an act of paying an extremely generous compliment to his guest. He says: 'me þin modsefa / licað leng swa wel, leofa Beowulf' (1853b–1854), 'your mind is more and more pleasing to me, dear Beowulf', attributing qualities to the mind inferred by observable

[95] For a translation of this text, collected within the *Acta Sanctorum*, see Michael J. B. Allen and Daniel G. Calder, *Sources and Analogues of Old English Poetry: The Major Latin Texts in Translation* (Cambridge, 1976), pp. 60–8.

behaviour – here, specifically ideas communicated in discussion. What is remarkable about this comment within the formalised interaction between king and visiting hero is that Beowulf has just exercised his physical rather than mental prowess, in that he has just performed the brave and miraculous feat of ridding the environs of two supernaturally strong monsters. Beowulf has communicated his ideas and values, though, both in speech and in observable action, so Hrothgar has logically had a reasonable opportunity for gaining the other man's mental measure. The text therefore implies the magnanimous king's possession of superior mind-reading skill, not just his understandable gratitude for heroic feats.

So, even though Hrothgar's extravagant compliment seems to have been crafted according to prevailing heroic ideals and the rhetoric of literary praise, it also encodes a socially embedded view about the mind and mind-reading. However we translate 'modsefa' in this citation ('mind', 'character', 'self', or 'disposition'), Hrothgar is claiming to have access to Beowulf's essential self by means of his observable behaviour: he is using his Theory of Mind on which to base an evaluation of Beowulf's character. In turn, the poet responsible for crafting these lines reflects not just assumptions about the mind-reading capacities of his audience in being able to interpret these lines, but more precisely in his representation of this scene between Hrothgar and Beowulf the cultural valorisation of acute mind-reading ability. As with St Helena, mental fortitude is matched with mental perspicacity as an attribute of an exceptional individual.

Grendel, the humanoid monster and descendant of Cain, on the other hand, lacks many human qualities, including empathy. This limitation is similar to autistic mindblindness, or reduced Theory of Mind, that limits empathetic responses.[96] Grendel is presented as a being who can imagine that others have minds, but cannot feel compassion or moral inhibition. He thereby demonstrates limited Theory of Mind in action, in that he can interpret and predict behaviour, but perceives other human beings as prey or food. He does demonstrate some cognitive functioning: he is able to feel emotion and to make inferences. When we are first introduced to Grendel, we are told that he 'þrage geþolode / . . . dream gehyrde / hludne in healle' (87a–9a), 'endured the torment of hearing rejoicing in the hall'. Grendel is represented as a being who can experience suffering, and is also able to experience jealousy and curiosity. The text presents him ambivalently as a monster bent on his prey and also a cold-hearted murderer whose intentions are communicated to the reader. Grendel travels to Heorot in order to see how the warriors had settled down for the night after their drinking (115–17), and later returns, after Beowulf's arrival there, intending slaughter: 'mynte se manscaða manne cynnes / sumne besyrwan' (712–713a), 'the murderer intended to snare someone'. He later sums up Beowulf's superior strength and decides to leave the fight: 'mynte se mæra . . . fleon on fenhopu'

96 On this condition, see Simon Baren-Cohen, 'Theories of the Autistic Mind', *Psychologist* 21.2 (2008), 112–16, and M. H. Davis, *Empathy: A Social Psychological Approach* (Boulder, 1994).

(762a–764a), 'the notorious one intended . . . to flee to the fens'. The emphasis placed on his intentions positions Grendel as a creature with recognisably human mental processing, a characteristic that allows his brutal murder of many people and disregard for human suffering to appear all the more monstrous by contrast when he treats the human body as food and intends to kill everyone in sight (730b–734a):

Þa his mod ahlog;
mynte þæt he gedælde, ær þon dæg cwome,
atol aglæca anra gehwylces
lif wið lice, þa him alumpen wæs
wistfylle wen.

Then his mind rejoiced; the terrible monster intended that before the day came, he would have separated life from limb of each of them, since the expectation of a fill of feasting had arisen for him.

The access to the mind of a creature who cannot imagine the minds, thoughts and feelings of others allows his cognitive deficiency to be made obvious. Similarly, the monster's inadequate imagination is highlighted when the reader is invited to see the working Theory of Mind in others, especially Beowulf, but also in a distributed way via songs, among the community that Grendel ravages and beyond, and to perceive that Grendel did not want to negotiate in the normal manner (149b–152a):

Forðam [gesyne] wearð
ylda bearnum, undyrne cuð
gyddum geomore þætte Grendel wan
hwile wið Hroþgar.

Therefore it became evident, clearly known, sadly in songs, to the sons of men that Grendel had long waged war against Hrothgar.

Grendel, then might very well be modelled on monstrous and evil human archetypes, as has been well demonstrated,[97] but the specifics of his antisocial behaviour seem to have been drawn from the cognitive impairment now known as mindblindness. This deficiency manifests in lacking or reduced abilities in terms of communication, imagination and social interaction, all characteristics demonstrated by Grendel.[98] By implication, he is lacking a definitive human attribute, empathy. He is shown to possess the cognitive aspect of empathy – some vestigial mind-reading ability – but not the affective response (identified as affective empathy[99]). A cognitive approach to the way in which the creature has been characterised and is seen to behave would allow access to culturally embedded views about mind-reading, emotion, intention, empathy, morality, guilt and many other things. More broadly, further analysis of contextually specific ideas about definitively

97 See the discussion and list of references in *Klaeber's Beowulf*, ed. Fulk, Bjork and Niles, p. xliv.
98 See Baren-Cohen, 'Theories of the Autistic Mind', p. 112.
99 Baren-Cohen, 'Theories of the Autistic Mind', p. 114, citing Davis, *Empathy*, 1994.

human mind-reading abilities demonstrated in this and other texts would helpfully progress our understanding of the constitutive power of culture on cognition in Anglo-Saxon England.

The cultural process of cognition in Anglo-Saxon England

Beyond these individual texts, and their specific instances of mind-reading, distributed cognition and the representation of mental processes within this literary culture, is the broader question of what the form and content of Old English poetic texts can tell us about the cultural process of cognition in Anglo-Saxon England: how cultural practices can have an impact on cognitive functioning, and how this negotiation can be encoded in literary texts. If we assume, as many do, that Old English verse was composed for oral recitation, then it follows that texts such as these that feature complex embedded mental states and multiple levels of intentionality make strenuous demands on their audiences – demands perhaps mitigated by textual and conceptual familiarity, the associations forged by alliteration and repetitive patterning and 'appositive style', and other literary strategies. Nevertheless, the evidence of the cognitive complexity of the texts, as well as the demonstrated ways in which cognition and culture interact, points to literary encounters that look very familiar to us today, that rely on the same cognitive architecture and mental processes, and that deploy language in a manner very much as our own contemporary poetry does. Ideally, cognitive approaches will be combined in future with literary and stylistic analyses of Old English texts, in order to develop these lines of inquiry further, and to consider more fully how cognitive processes determine literary structure and interpretation in temporally remote cultures.

These instances give some idea of how Cognitive Science can facilitate new ways of understanding the dynamic of didactic poetry, and Anglo-Saxon poetry in turn can demonstrate that mind-reading is not wholly the province of the modern novel, but rather has characterised literary texts for much longer, and probably can be found in all texts, whether composed scribally or for oral performance. Indeed, the cognitive and cultural inter-reliance of narrative, imagination and complex embedded mental states from several different lines of research inquiry suggests just how deep and pervasive the mutual impact appears to be, and how much more research is required to understand these relationships more fully. But the brief indications above of specific manifestations of the capacity for extrapolating thoughts and feelings from behaviour in *The Dream of the Rood, Elene* and *Beowulf* show what cognitive approaches can do for Old English and in turn what Old English can bring to the cognitive study of literature and culture.

6

Anglo-Saxon Perspectives on Autobiographical Memory and the Self

The cognitive approaches outlined in the foregoing chapters all have their own field-specific bases, alignments with particular professional associations, key scholarly figures and characteristic modes of inquiry. They all have something to offer the Anglo-Saxonist, both in terms of specific modes of textual engagement and lines of inquiry, but also in promoting and expanding the cognitive turn in Literary Studies generally through combination. Beyond their emerging disciplinary developments and distinctive ways of doing the business of interpretation, their chief value is in foregrounding the complexity and interrelatedness of the functioning of the evolved human brain and its cultural context. This dynamic relationship is in play in all cognitive approaches, but sits at the core of Cognitive Cultural Studies, which is why this particular field seems to be the most attractive and productive for Anglo-Saxon Studies.

But Cognitive Cultural Studies has yet to develop fully the range and potential of the fruitful embrace of Cognitive Science and Literary/Cultural Studies – which itself is not surprising, given the huge and voracious scope of the former. So, there is a considerable amount of inquiry and exploration left to be done, a task to which the Anglo-Saxonist can contribute, not only for the sake of the vigour and prosperity of field itself, but also to ensure that the cognitive turn takes cognisance of the full diachronic sweep of Literary Studies and engages with historical depth right back to the earliest forms of English. The methodological relevance of contemporary science is not the sole province of recent literature; and legitimate arguments about the interplay of cognition and culture can only gain as a result of broadening the scope of inquiry beyond literature of the modern era. Similarly, Cognitive Science cannot only permit, but can actively embrace, ideas and methodologies from the literary scholar working on Old English texts.

Given these considerations, this chapter and the next set out to demonstrate the contribution that can be made by Anglo-Saxon Studies, not only to emerging cognitive approaches to literature and culture, but to Cognitive Science itself. The first section of this chapter explores the reliance of Cognitive Science on ideas of narrative, especially in relation to memory and the development of the self. It then turns to the potential of a fuller collaboration between this field with Literary Studies in general, and Anglo-Saxon Studies in particular, in the consideration of the capacity of early poetry to represent subjective states and autobiographical memory. This representation not only enacts the dynamic interaction of culture and cognition, but also demonstrates the long diachronic sweep of literary self-conceptualisation and the long-term consistency of cognitively informed literary structures and responses.

Narrative and autobiographical memory

One phenomenon not fully addressed in cognitive approaches is the degree to which Cognitive Science has embraced literary tools and terminology, particularly in relation to the narrative quality of autobiographical memory, but also the importance of narrative organisation for the development and maintenance of a sense of self or durable individual identity. Scholars in Psychology, Philosophy and other areas of Cognitive Science have approached this inter-relationship of identity and story, and in turn have embraced the literary definition of narrative as a representation of a causally related series of events,[1] – with the emphasis on its organised, sequential properties – and have deployed this idea widely in studies of human memory. Of the many types of memory now recognised, 'autobiographical memory' is understood here as the recollection of one's own prior experiences, a reconstructive act that is always culturally situated, context sensitive and susceptible to narrative configuration. This understanding combines the view that autobiographical memories are 'transitory dynamic mental constructions generated from an underlying knowledge base',[2] with Erll and Nünning's conception of cultural memory,[3] and Bruner's influential idea that all human experience is shaped and organised as narrative, out of a cultural context of shared stories.[4]

The narrative quality of memory in particular has been endorsed and examined by many scholars in both Psychology and Philosophy.[5] Similarly, the inter-relationship of narrative and the formation of personal identity, or the self, is well established and continues to attract theoretical and

1 See the definitions in H. Porter Abbot, *The Cambridge Introduction to Narrative*, 2nd edn (Cambridge, 2008), pp. 13–5; and Mieke Bal, *Narratology: Introduction to the Theory of Narrative*, 2nd edn (Toronto, 1997), pp. 3–10.

2 Martin A. Conway and Christopher W. Pleydell-Pearce, 'The Construction of Autobiographical Memories in the Self-Memory System', *Psychological Review* 107.2 (2000), 261–88, at p. 261.

3 Astrid Erll and Ansgar Nünning, eds, *Cultural Memory Studies: An International and Interdisciplinary Handbook* (Berlin and New York, 2008).

4 Jerome S. Bruner, 'The Narrative Construction of Reality', *Critical Inquiry* 18 (1991), 1–21; 'Self-Making and World-Making', in *Narrative and Identity: Studies in Autobiography, Self and Culture*, ed. J. Brockmeier and D. Carbaugh (Amsterdam and Philadelphia, 2001), pp. 25–37; *Making Stories: Law, Literature, Life* (New York, 2002).

5 K. Nelson, 'Narrative and the Emergence of a Consciousness of Self', in *Narrative and Consciousness: Literature, Psychology and the Brain*, ed. G. D. Fireman, Ted E. McVay, Jr, and Owen J. Flanagan (Oxford and New York, 2003), pp. 17–36; Nelson, 'Narrative and Self, Myth and Memory: Emergence of the Cultural Self', in *Autobiographical Memory and the Construction of a Narrative Self: Developmental and Cultural Perspectives*, ed. R. Fivush and C. A. Haden (Mahwah, NJ, 2003), pp. 3–28; David Herman, 'Stories as a Tool for Thinking', in *Narrative Theory and the Cognitive Sciences* (Stanford, 2003), pp. 163–92; and J. David Velleman, *Self to Self: Selected Essays* (Cambridge, 2006), esp. Chapter 9, 'The Self as Narrator', pp. 203–23.

empirical study.[6] The very act of putting story into language has a bearing on the qualities and framing of that story, especially when the self is the subject of that narrative. As Bruner argues, 'autobiography is life construction through "text" construction'.[7] This specifically narrative and indeed discursive understanding of the self has been embraced by many scholars, and developed in several directions,[8] including those that are linguistically focused. Indeed, Olney goes beyond the text, to figurative language as a source of self, arguing that autobiography operates within the process of self-expression, and therefore, self-design, through the use of dominant metaphors.[9] Like Olney, Eakin argues explicitly that memory shapes the past through the motivated agendas of the autobiographer, engaged in a textualising process.[10] Key to this conception is the notion of representation in both self-conception and textuality. For instance, Velleman defines the self as a 'person's own reflexive representations, which make up his self-image or self-conception'.[11]

In all, there is growing psychological research on the precise relationships between narrative and autobiographical memory, an interplay that is observed from the literary side of the fence too. For instance, Kay Young and Jeffrey Saver argue that 'what predominates or fundamentally constitutes our consciousness is the understanding of self and world in story'.[12] Conversely, Cognitive Scientists have begun to use literary data in studies of autobiographical memory. Building on the work of Anderson and Conway on the structure, storage and retrieval of autobiographical memories, Radvansky and his collaborators have demonstrated the correlations between the compre-

6 D. Dennett, *Consciousness Explained* (London, 1991); R. Fivush, 'Owning Experience: Developing Subjective Perspectives in Autobiographical Narratives', in *The Self in Time: Developmental Perspectives*, ed. C. Moore and K. Lemmon (Mahwah, NJ, 2001); A. Thorne, 'Personal Memory Telling and Personality Development', *Personality and Social Psychology Review* 4 (2000), 45–56; Margaret H. Freeman, 'From Substance to Story: Narrative, Identity, and the Reconstruction of the Self', in *Narrative and Identity*, ed. Brockmeier and Carbaugh, pp. 283–98; J. Brockmeier and R. Harré, 'Narrative: Problems and Promises of an Alternative Paradigm', in *Narrative and Identity*, ed. Brockmeier and Carbaugh, pp. 59–73; Velleman, *Self to Self*; D. D. Hutto, *Folk Psychological Narratives: The Socio-Cultural Basis of Understanding Reasons* (Cambridge, MA, 2008); K. C. McLean, M. Pasupathi and J. L. Pals, 'Selves Creating Stories Creating Selves: A Process Model of Self-development', *Personality and Social Psychology Review* 11 (2007), 262–78. An alternative view, that a sense of self arises from embodied experience, specifically the 'interiorization of speech', is argued by R. Menary in 'Embodied Narratives', *Journal of Consciousness* 15 (2008), 63–84.

7 Bruner, 'The Autobiographical Process', in *The Culture of Autobiography: Constructions of Self-Representation*, ed. R. Folkenflik (Stanford, 1993), pp. 38–56, at p. 55.

8 See, for example, D. P. McAdams, *The Redemptive Self: Stories Americans Live By* (Oxford, 2006); D. P. McAdams, R. Josselson and A. Lieblich, eds, *Identity and Story: Creating Self in Narrative* (Washington, DC, 2006); J. McCarthy, *Dennett and Ricoeur on the Narrative Self* (Amherst, NY, 2007); and R. Josselin and A. Lieblich, 'Reflections on the Narrative Study of Lives', *Narrative Inquiry* 19 (2009), 183–98.

9 J. Olney, *Metaphors of Self: The Meaning of Autobiography* (Princeton, 1972).

10 P. J. Eakin, *Fictions in Autobiography: Studies in the Art of Self Invention* (Princeton, 1985); and *Living Autobiographically: How We Create Identity in Narrative* (Ithaca and London, 2008).

11 Velleman, *Self to Self*, p. 355.

12 Kay Young and Jeffrey Saver, 'The Neurology of Narrative', *SubStance* 30 (2001), 72–84, at p. 73.

hension and retrieval of details from narrative fiction and autobiographical memories.[13]

Outside Literary Studies, published autobiography is frequently the subject of analysis, because it provides a written account of identity formation[14] and on psychological processes.[15] But the interplay of context and subject has occluded the textual component in these studies. For instance, the textual transmission of generic frames, and in turn, ways of conceiving of autobiographical memory and life narratives as autobiographical are yet to be explored, as is the durable impact of ancient and medieval life writing. There remains work to be done, then, on the impact of these published texts as part of the culturally embedded set of narrative templates which underpin self-formation and the creation and articulation of individual life narratives, both individually and collaboratively.[16] In particular, Cognitive Science is yet to consider, and could explore in an interdisciplinary manner with Literary Studies, the influence of the culturally transmitted generic structures and narrative schemas of life-writing, many of which can be traced back to texts from much earlier cultures, including Anglo-Saxon England. These schemas participate in the social distribution of ways of remembering, and of organising and recounting those memories. Similarly, the capacity of poetry to represent subjective states – to provide the medium whereby the reader may explore what it is to try on another consciousness – provides unique access to inner lives, to autobiographical memory and to an individual, yet in part culturally determined sense of self.[17] Old English poetry enacts this capacity, so can be used as valuable data for any consideration of the diachronic range of the interplay among poetic expression, cultural transmission, subjectivity and autobiographical memory.

The interests and omissions of contemporary Psychology, then, suggest the genuine potential of a cross-disciplinary partnership with Literary Studies, particularly given the huge amount of scholarship on the ancient, durable and highly influential art of life-writing. Modern autobiographies can trace their sources of influence as far back as Augustine's *Confessions* from 398 CE, the text that instituted 'the inwardness of radical reflexivity',[18] and had a profound effect on the shape, language and presentation of self-examination and in turn, on the creation of the idea of the self through the narrative organisation of and interpretation of personal memories. This phenomenon

13 S. J. Anderson and M. A. Conway, 'Investigating the Structure of Autobiographical Memories', *Journal of Experimental Psychology: Learning, Memory and Cognition* 19 (1993), 1178–96; G. A. Radvansky, D. E. Copeland and R. A. Zwaan, 'A Novel Study: Investigating the Structure of Narrative and Autobiographical Memories', *Memory* 13 (2005), 796–814. Interestingly, this study uses a fictionalised autobiography, Carol Shields's *The Stone Diaries*, as a case study.

14 For example Eakin, *Fictions in Autobiography* and *Living Autobiographically*.

15 J. A. Belzen and A. Geels, eds, *Autobiography and the Psychological Study of Religious Lives* (Amsterdam and New York, 2008).

16 See Antonina Harbus, 'Exposure to Life-Writing as an Impact on Autobiographical Memory', *Memory Studies* 20 (2011), 1–15.

17 Antonina Harbus, 'The Medieval Concept of the Self in Anglo-Saxon England', *Self and Identity* 1.1 (2002), 77–97.

18 Charles Taylor, *Sources of the Self: The Making of Modern Identity* (Cambridge, 1989), p. 131.

of self-formation through the act of self-representation in literary discourses, predates the time to which many scholars ascribe the flourishing of the art, the early-modern period;[19] evidence for it can be found in the poetry of Anglo-Saxon England, as well as in other medieval literatures, and in classical texts.

These ideas have to be considered in conjunction with the arguments from Cognitive Science that culture has an impact on the modes and qualities of cognition,[20] and that culture in large part determines the models of selfhood on which individual selves are developed.[21] Many psychologists insist on the social and cultural, as well as the cognitive, impact on identity, and in turn, the influence of culturally significant narratives on collective memory,[22] and identity formation.[23] An ethnographic generation of master narratives provides shared material deployed in individual self-formation and autobiographical memory.[24] To trace these inter-relationships, the following discussion will consider theoretical views on the constitution of and influences on self; literature as a source of information on these subjects; and evidence from Old English poetry to support the contentions that the medieval English literary construction of the self is both familiar and durable.

The self: concept and context

Cognitive scientists, while still puzzling over the question 'How does the brain create a self?',[25] widely agree that the self is composite, variable, unstable

19 R. Porter, ed., *Rewriting the Self: Histories from the Renaissance to the Present* (London, 1997); M. Mascuch, *Origins of the Individualist Self: Autobiography and Self-Identity in England, 1591–1791* (Cambridge, 1997); and R. Bedford, L. Davis and P. Kelly, eds, *Early Modern Autobiography: Theories, Genres, Practices* (Ann Arbor, 2006). The continued fascination of literary scholars with the genre is evidenced by the recent four-volume collection of important critical articles on the subject. See T. L. Broughton, ed., *Autobiography* (Abingdon, 2007).

20 R. G. D'Andrade, 'Cultural Meaning Systems', in *Culture Theory: Essays on Mind, Self, and Emotion*, ed. R. A. Shweder and R. A. LeVine (Cambridge, 1984); M. Cole, *Cultural Psychology: A Once and Future Discipline* (Cambridge, MA, 1996); and S. Kitayama and H. Markus, eds, *Handbook of Cultural Psychology* (New York, 2007). See also Chapter 5 above and Chapter 7 below.

21 Bruner, *Making Stories*.

22 E. Reese and R. Fivush, 'The Development of Collective Remembering', *Memory* 16 (2008), 201–12; Amanda J. Barnier, John Sutton, Celia B. Harris and Robert A. Wilson, 'A Conceptual and Empirical Framework for the Social Distribution of Cognition: The Case of Memory', *Cognitive Systems Research* 9 (2008), 33–51; James V. Wertsch, 'Collective Memory', in *Memory in Mind and Culture*, ed. Pascal Boyer and James V. Wertsch. (Cambridge, 2009), pp. 117–37; and H. L. Roediger, F. M. Zaromb and A. C. Butler, 'The Role of Repeated Retrieval in Shaping Collective Memory', in *Memory in Mind and Culture*, ed. Boyer and Wertsch, pp. 138–71.

23 P. L. Hammack, 'Narrative and the Cultural Psychology of Identity', *Personality and Social Psychology Review* 12 (2008), 222–47; and Michael Bamberg, 'Who am I? Narration and its Contribution to Self and Identity', *Theory and Psychology* 21.1 (2011), 3–24.

24 James Wertsch, *Mind as Action* (Oxford, 1998); A. Thorne and K. C. McLean, 'Telling Traumatic Events in Adolescence: A Study of Master Narrative Positioning', in *Autobiographical Memory and the Construction of a Narrative Self: Developmental and Cultural Perspectives* (Mahwah, NJ, 2003), pp. 169–86; D. P. McAdams, R. Josselson and A. Lieblich, eds, *Identity and Story*; and H. L. Williams and M. A. Conway, 'Networks of Autobiographical Memories', in *Memory in Mind and Culture*, ed. P. Boyer and J. V. Wertsch (Cambridge, 2009), pp. 33–61.

25 Todd E. Feinberg, 'The Nested Neural Hierarchy of the Self', *Consciousness and Cognition* 20.1 (2011), 4–15, at p. 4.

and dynamic, even dialogic. It is 'located at the centre of a nexus of interrelated and partly overlapping concepts describing personhood', blending at its outer edges with personality and character.[26] The self is currently a matter of wide-ranging interest to Cognitive Science, as the range of entries in a recent handbook attests.[27] It has variously been conceived as a form of 'understanding' that is reflexive and context-sensitive;[28] as a 'process of the brain';[29] and most popularly, following Bruner, as a specifically narrative process: 'we somehow create the self by narratively interpreting a subset of our life events'.[30] Indeed, it is apparent that this focus on narrative, and on the context in which the self operates dynamically, has become a dominant interest in Psychology.[31] Notwithstanding the breadth and depth of research into this elusive topic, it is unfortunate that studies of the self in Cognitive Science, even in Narrative Psychology, do not consider historical developments and transmissions beyond a single human life span, and barely engage with literary concepts and methods, even when considering the schemas and templates on which life stories are modelled. As Peter Raggatt explains:

> The problem of identity can be approached by allowing for multiplicity and conflict in the self. The approach described offers some promise as a generative paradigm for psychology, particularly for narrative psychology and the study-of-lives approach. Yet the discipline has largely left these ideas to the novelists, biographers, and artists, perhaps because it is so difficult to study a human subject that is shifting positions, caught between conflicting stories.[32]

Beyond this limitation are other field-specific restrictions that inhibit a full consideration of the dialogic nature of the self and its indebtedness to transmitted models: just as psychologists are unlikely to consider medieval texts as sources of evidence for earlier views on the self, so too are medievalists who examine the self as a topic unlikely to engage with Cognitive Science.[33] After examining discussion from both sides, the potential offered

26 Kristján Kristjánsson, *The Self and its Emotions* (Cambridge, 2010), p. 26.

27 Shaun Gallagher, *The Oxford Handbook of the Self* (Oxford, 2011): see the 'Introduction', pp. 1–29, esp. p. 27, for a discussion on the complexity and variety of approaches to the self.

28 Jack Martin, Jeff H. Sugarman and Sarah Hickinbottom, *Persons: Understanding Psychological Selfhood and Agency* (New York, 2010), pp. 36–7.

29 Todd E. Feinberg, *From Axons to Identity: Neurological Explorations of the Nature of the Self* (New York and London, 2009), p. 209. See also Hubert J. M. Hermans, 'The Dialogical Self: A Process of Positioning in Space and Time', in *The Oxford Handbook of the Self*, ed. Gallagher, pp. 654–80.

30 Valerie Gray Hardcastle, *Constructing the Self*, Advances in Consciousness Research 73 (Amsterdam, 2008), p. 18. See also D. Dennett, 'The Self as a Center of Narrative Gravity', in *Self and Consciousness: Multiple Perspectives*, ed. F. S. Kessel, P. M. Cole and D. L. Johnson (Hillsdale, NJ, 1992), pp. 103–15; Velleman, 'The Self as Narrator'; Dan Hutto, *Folk-Psychological Narratives: The Socio-Cultural Basis of Understanding Reasons* (Cambridge, MA, 2008); and Marya Schechtman, 'The Narrative Self', in *Oxford Handbook of the Self*, ed. Gallagher, pp. 394–416.

31 See, for example, S. Kirschner and J. Martin, eds, *The Sociocultural Turn in Psychology: The Contextual Emergence of Mind and Self* (New York, 2010).

32 Peter T. F. Raggatt, 'Multiplicity and Conflict in the Dialogical Self: A Life-Narrative Approach', in *Identity and Story*, ed. McAdams *et al.*, pp. 15–35, at p. 33.

33 See, for instance, A. C. Spearing, *Textual Subjectivity: The Encoding of Subjectivity in Medieval Narratives and Lyrics* (Oxford, 2005), who chooses not to engage with Cognitive Science when he challenges the idea that a narrator gives voice to subjectivity in medieval texts, which

by a genuinely cognitive cross-disciplinary approach can be exemplified by excerpts from Old English elegiac texts.

That concepts of selfhood vary across cultures is now a widely accepted idea within Psychology, Anthropology, Philosophy and other fields. Psychologists have investigated the impact of the specific construction of self on social behaviour,[34] and on cognitive, emotional and motivational experience and behaviour,[35] as well as culturally determined self construction.[36] Anthropologists have likewise demonstrated that a sense of self is heavily culturally determined,[37] as have philosophers,[38] and linguists, particularly those engaged in cross-cultural research.[39] Many writers, from a range of disciplines, conflate the self with identity, consciousness and individuality, which itself can be a source of confusion and inaccuracy.[40] Despite the advancement of these theories, it has not been customary to compare historical variations on concepts of selfhood in order to investigate further the contingencies of contemporary views, except in some brief studies.[41] Even in a very recent overview article,

cannot be read as 'the expression of human consciousness' (p. 2), and rather argues that a distributed form of subjectivity is discernible through deictic references.

34 Harry C. Triandis, 'The Self and Social Behaviour in Differing Cultural Contexts', *Psychological Review* 96. 3(1989), 506–20.

35 Hazel Rose Markus and Shinobu Kitayama, 'Culture and the Self: Implications for Cognition, Emotion, and Motivation', *Psychological Review* 989.2 (1991), 224–53.

36 The work of Kenneth J. Gergen is prominent in this field: see *Relational Being: Beyond Self and Community* (Oxford, 2009), and his overview article, 'The Social Construction of Self', in *The Oxford Handbook of the Self*, ed. Gallagher, pp. 633–53. See also: Susan E. Cross, 'Culture, the Self, and Identity: Introduction to the Special Issue', *Self and Identity* 2 (2003), 265–8; Y. Kashima, E. Kashima, T. Farsides, K. Uichol, F. Strack, L. Werth, Lioba and M. Yuki, 'Culture and Context-Sensitive Self: The Amount and Meaning of Context-Sensitivity of Phenomenal Self Differ across Cultures', *Self and Identity* 3 (2004), 125–41; and Ching Wan, K. Dach-Gruschow, S. No and Y. Hong, 'Self-Definitional Functions of Culture', in *Cultural Processes: A Social Psychological Perspective*, ed. Angela K. Leung, Chi-yue Chiu and Ying-yi Hong (Cambridge, 2011), pp. 111–35.

37 Dorothy Holland, 'Selves as Cultured: As Told by an Anthropologist who Lacks a Soul', in *Self and Identity: Fundamental Issues*, ed. Richard D. Ashmore and Lee Jussim (New York and Oxford, 1997), pp. 160–90; A. J. Marsella, G. DeVos, and F. L. K. Hsu, eds., *Culture and Self* (New York, 1985); and Richard A. Shweder, 'Properties of Culture: An Ethnographic View', *Culture Theory*, ed. Shweder and LeVine, pp. 67–87.

38 For example, Richard Sorabji, *Self: Ancient and Modern Insights about Individuality, Life, and Death* (Chicago, 2006).

39 For example, Edward Slingerland, 'Conceptions of the Self in the Zhuangzi: Conceptual Metaphor Analysis and Comparative Thought', *Philosophy East and West* 54.3 (2004), 322–4.

40 See, for example, the pleas for precision in the use of the terms in Mark R. Leary, Editorial: 'What is the Self? A Plea for Clarity', *Self and Identity* 3.1 (2004), 1–3.

41 John Adamopoulos, 'Analysis of Interpersonal Structures in Literary Works of Three Historical Periods', *Journal of Cross-Cultural Psychology* 13.2 (1982), 157–68; John Adamopoulos and Robert N. Bontempo, 'Diachronic Universals in Interpersonal Structures: Evidence from Literary Sources', *Journal of Cross-Cultural Psychology* 17.2 (1982), 169–89; Roy F. Baumeister, 'How the Self Became a Problem: A Psychological Review of Historical Research', *Psychological Review* 52.1 (1987), 163–76; Baumeister,'The Self and Society: Changes, Problems, and Opportunities', in *Self and Identity: Fundamental Issues*, ed. Richard D. Ashmore and Lee Jussim (New York and Oxford, 1997), pp. 191–217 ; Kurt Danziger, 'The Historical Formation of Selves', in *Self and Identity: Fundamental Issues*, ed. Richard D. Ashmore and Lee Jussim (New York and Oxford, 1997), pp. 137–59; and Mark Elvin, 'Between the Earth and Heaven: Conceptions of the Self in China', in *The Category of the Person*, ed. Michael Carrithers, Steven Collins and Steven Lukes (Cambridge, 1985), pp. 156–89.

the narrative skips from classical ideas, briskly past Augustine, to the early-modern period, with only a paragraph on the Middle Ages, and then only to refer to the reception and transmission of classical ideas.[42] Cultural variation is not routinely a matter of consideration. Indeed, it is only relatively recently that anthropologists and psychologists have argued that 'a complete account of self and identity requires attention to cultural and historical forces as an integral component', in a 'dynamic construction' model of self-construction that is interdependent, not independent.[43]

Part of that socio-historical context is provided by the production, consumption and transmission of literary texts, genres and schemas, a complete exploration of which requires diachronic as well as synchronic investigations. Specifically, medieval literary evidence can be used to show that concepts of selfhood vary across time and that poetic texts in particular are useful sources of information for recovering formulations of the self from earlier cultures. Scholars have begun to pursue this line of inquiry, with some useful results, but a collaboration with Cognitive Science can take the exploration further, to distinguish apparent consistencies in English constructions of the self from those concepts which are more temporally, as well as culturally, specific. Indeed, more detailed engagement with the concepts and language relating to different facets and qualities of the self, taken from Cognitive Science, can provide the tools for a more nuanced understanding of textual-cognitive interaction in self formation and expression, particularly in relation to autobiographical memory.

There are other methodological matters to consider as well. It is now apparent that understandings of the self that are based on contemporary, Western concepts of the individual and human nature might be considered unduly restricted, perhaps even 'monocultural'.[44] Starting at the basic level of semantics, it seems that many current premises relating to the self are predicated on an extremely culturally specific construction of selfhood (based on contemporary, English-speaking discourse on the self), not only because the English word *self* is inescapably 'lexically embodied in English',[45] but also because the word semantically carries connotations ascribed by *contemporary* English-speaking cultures. Apart from the restriction to Western, English-speaking cultures, the prevalent model of the individual is temporally bound to modern connotations of the word *self*. Even though our English word *self* can be traced to Old English texts (specifically, those written down around the ninth century, occurring as an adjective or pronoun), it does not acquire the contemporary substantive connotation, 'individual identity' or 'ego' until much later, and now has a wide semantic field in both popular and

42 John Barresi and Raymond Martin, 'History as Prologue: Western Theories of the Self', in *The Oxford Handbook of the Self*, ed. Gallagher, pp. 33–56

43 Glenn Adams and Vivan Afi Dzokoto, 'Self and Identity in African Studies', *Self and Identity* 2 (2003), 345–59, at p. 356.

44 Markus and Kitayama, 'Culture and the Self'.

45 Anna Wierzbicka, 'A Conceptual Basis for Cultural Psychology', *Ethos* 21.2 (1993), 205–31, at p. 209; and Dorinne K. Kondo, *Crafting Selves: Power, Gender and Discourses in a Japanese Workplace* (Chicago, 1990).

technical discourse.[46] The *Oxford English Dictionary* ascribes the earliest usage of the word *self*, meaning 'a permanent subject of successive and varying states of consciousness', to the late-seventeenth century (s.v. *self*, C.sb.3). In accordance with this semantic fact, the philosophical identification of the self as an entity is usually considered to be a relatively recent phenomenon and the emergence of the self as an object of study to be even more recent, traceable to the 1890s.[47]

If we look beyond the restrictive and linguistically specific formation of the self confined to present-day usage, and indeed look beyond the use of the word altogether, focusing instead on expressions of the more universal notion of individual identity and agency in conjunction with expressions of folk psychologies,[48] we have access to a much wider recognition of the individual self. If it is possible to discern a change in the 'folk model of the mind' within a single culture over time,[49] a phenomenon encoded in the changing connotations of individual terms such as *mind* and *self*,[50] it seems inappropriate to apply our contemporary vocabulary and semantic connotations to earlier folk constructions of identity. Broadening the inquiry to historical variation in the consciousness of self, we can acquire some insight into our own culturally composed fundamentals of self. In this manner, some scholars working within Cognitive Science have looked to literary texts produced in earlier societies for evidence of cultural models of mental life, moral agency and interpersonal relations,[51] just as others have sought the same goals in tribal or 'non-Western' societies.[52]

In the humanities, the most sustained investigations into the ways in which the self has been considered over time have been in the field of History. Many have proposed that introspection is a phenomenon of the early-modern and modern periods, and correlates with the rise of the novel as a genre. Indeed, even those working outside Literary Studies rely on this orthodoxy in order to drive home their arguments, and literary texts are often used as evidence, without consideration of their status as a source of data. For instance, Charles Taylor, a moral philosopher, cites a literary text, Conrad's *Heart of Darkness*, as a cultural reference point by which to illustrate this point, and goes on to link the rise in psychological awareness in England from the seventeenth

46 *Oxford English Dictionary*, 2nd edn (Oxford, 1989), s.v. 'self'. In Old English, forms of *self/selfa* were used as an adjective or pronoun in the sense 'itself, himself, herself', or in conjunction with a pronoun for emphasis, as in 'I, myself'. For the full history, see Elly van Gelderen, *A History of English Reflexive Pronouns: Person, Self, and Interpretability*, Linguistik Aktuell 39 (Amsterdam and Philadelphia, 2000).

47 Richard D. Ashmore and Lee Jussim, eds, *Self and Identity: Fundamental Issues* (New York and Oxford, 1997); Frank Johnson, 'The Western Concept of the Self', in *Culture and Self*, ed. Marsella, DeVos and Hsupp, pp. 91–138.

48 Wierzbicka, 'A Conceptual Basis for Cultural Psychology'.

49 D'Andrade, 'Cultural Meaning Systems', p. 138.

50 Anna Wierzbicka, *Semantics, Culture and Cognition: Universal Human Concepts in Culture-Specific Configurations* (New York, 1992).

51 Elvin, 'Between the Earth and Heaven'; Adamopoulos and Bontempo, 'Diachronic Universals in Interpersonal Structures'.

52 Steven M. Parish, 'The Sacred Mind: Newar Cultural Representations of Mental Life and the Production of Moral Consciousness', *Ethos* 19.3 (1991), 313–51.

century to the rise of the practice of recording personal feelings in diaries and the rise of the novel as a vehicle for representing introspection.[53] Danziger similarly cites the 'cultural ambience of Romanticism' as the trigger for the development of 'retrospective self-examination'.[54] It is well documented that a poet, not a scientist – Coleridge – brought the word *psychology* into England from Germany in the early 1800s, where it had been in use for a generation. After Charles Taylor's work on the self, there have been several studies from the perspective of Intellectual History, most notably Jerrold Seigel's *The Idea of The Self: Thought and Experience in Western Europe since the Seventeenth Century*.[55] Like Taylor, Seigel focuses on tracing the roots of modern ideas about the self, with an emphasis on early-modern intellectualism, a common focal point. For instance, the literary historian Stephen Greenblatt, like the psychologist Roy F. Baumeister, situates the rise of the consciousness of an inner self (construed as modern identity) to the sixteenth century,[56] though the date has been pushed back even further by historians and literary scholars. Several have striven to demonstrate that the idea of the individual was not a phenomenon of the Renaissance, as many contend, but rather can be traced to Scholasticism and the rise of vernacular texts in the Middle Ages,[57] and even further, to classical Greece.[58] The continuing debate on the subject has suggested that even this date might be challenged, so that, in a welcome change of direction, scholars have come to explore more fully the way in which subjectivity can be encoded in medieval texts. In particular, the notion that the literary personae of earlier texts were representative rather than unique has come to be criticised by cultural historians and scholars of earlier vernacular texts.[59]

53 Taylor, *Sources of the Self*, p. 184.

54 Kurt Danziger, *Marking the Mind: A History of Memory* (Cambridge, 2008), p. 105.

55 Jerrold Seigel, *The Idea of the Self: Thought and Experience in Western Europe since the Seventeenth Century* (Cambridge, 2005).

56 Stephen Greenblatt, *Renaissance Self-Fashioning: From More to Shakespeare* (Chicago, 1980); Baumeister, 'How the Self Became a Problem'.

57 Two of the most influential studies include Colin Morris, *The Discovery of the Individual, 1050–1200* (Toronto, 1972); and Carolyn Walker Bynam, *Jesus as Mother: Studies in the Spirituality of the High Middle Ages* (Berkeley, 1982), especially the chapter 'Did the Twelfth Century Discover the Individual?" See also John F. Benton, 'Consciousness of Self and Perceptions of Individuality', in *Renaissance and Renewal in the Twelfth Century*, ed. Robert L. Benson and Giles Constable (Cambridge, MA, 1982), pp. 263–95; and Sarah Spence, *Texts and the Self in the Twelfth Century* (Cambridge, 1996).

58 Raymond Martin and John Barresi, *The Rise and Fall of Soul and Self: An Intellectual History of Personal Identity* (New York, 2006).

59 Notably suggested some time ago by Bynam in *Jesus as Mother*; and developed more fully by others, including, for instance, Fiona Gameson and Richard Gameson, '*Wulf and Eadwacer, The Wife's Lament*, and the Discovery of the Individual in Old English Verse', in *Studies in Old English Language and Literature: 'Doubt Wisely': Papers in Honour of E. G. Stanley*, ed. M. J. Toswell and E. M. Tyler (London and New York, 1996), pp. 457–74; and the articles in *Image and Imagination of the Religious Self in Late Medieval and Early Modern Europe*, ed. Reindert Falkenburg, W. S. Melion and T. M. Richardson (Turnhout, 2007).

Medieval selves

That a sense of self might be recoverable from older literatures seems to be supported by attitudes expressed in the Social Sciences, and able to be expanded further through specifically cognitive approaches to older literatures. The prevailing view of cultural anthropologists is that a sense of self is constructed at least in part by the social environment, mediated by class, gender and personal factors, encoded in behaviour and recorded in texts.[60] Although both cultural relativism and some aspects of universalism are accepted in varying degrees by cognitive and social scientists, there is some consensus that the idea of the self is a universal attribute of the human condition,[61] whose nature is contingent upon a broad network of culturally specific ideas and the social situation of the individual.[62]

But medieval constructions of the self tend to be given only the briefest attention by psychologists, sociologists and anthropologists, even those exploring the history of the 'Western self',[63] a concept which itself has been subject to vigorous critiques.[64] When pre-modern conceptions of the self are examined, it is frequently for the purpose of demonstrating the developing biography of modern ideas about human identity and subjectivity, emerging social behaviour,[65] or our superior awareness of the self today. Often, descriptions of medieval conceptions of the self are extremely general as if all pre-modern cultures constructed the same sort of selves. Johnson captures all Western conceptions of the self from 'pre-Christian times until 1850' with this formulation: '(The self) was characterised by philosophic, theological and literary descriptions of the self (using the concepts of soul and mind) stressing the individual nature of subjective consciousness, the ontological separateness of both persons and things, and the tendency to locate responsibility for action in the individual.'[66]

Other evaluations of the medieval evidence are extremely cursory or otherwise dubious. Danziger considers that views of the self up to the seventeenth century were essentially negative because of the dependence upon

60 Adamopoulos and Bontempo, 'Diachronic Universals in Interpersonal Structures'; Danziger, 'The Historical Formation of Selves'; DeVos *et al.*, *Culture and Self*; Holland, 'Selves as Cultured'.

61 Irving Hallowell, *Culture and Experience* (Philadelphia, 1955); Michelle Z. Rosaldo, 'Toward an Anthropology of Self and Feeling', in *Culture Theory*, ed. Shweder and LeVine, pp. 136–57; Melford E. Spiro, 'Some Reflections on Cultural Determinism and Relativism with Special Reference to Emotion and Reason', in *Culture Theory*, ed. Shweder and LeVine, pp. 323–46; Wierzbicka, 'A Conceptual Basis for Cultural Psychology'.

62 Baumeister,'The Self and Society'; Holland, 'Selves as Cultured'; Richard A. Shweder, 'Properties of Culture: An Ethnographic View', in *Culture Theory*, ed. Shweder and LeVine, pp. 67–87.

63 For example, Martin *et al.*, *Persons*, pp. 12–13; Baumeister,'The Self and Society'; and Johnson, 'The Western Concept of the Self'.

64 Holland, 'Selves as Cultured'; Melford E. Spiro, 'Is the Western Conception of the Self "Peculiar" within the Context of World Cultures?', *Ethos* 21.2 (1993), 107–53.

65 Seigel, *The Idea of the Self*, addresses ancient and medieval ideas about the self as conflated with the soul and limited to religious views (p. 53). See also, Adamopoulos, 'Analysis of Interpersonal Structures'.

66 Johnson, 'The Western Concept of the Self', p. 128.

Christian constructions of the self as 'sinner';[67] and Baumeister, dating the rise of a sense of individuality to the period 1500–1800, argues that 'notions of selfhood in the Middle Ages in the West may have been far more collective than they are now'.[68] No evidence is brought to support this unlikely claim. Similarly, his comment that 'late medieval awareness of selfhood was crude by modern standards, suggesting that self-knowledge was not regarded as an important problem'[69] is supported only by reference to the second-hand and very dated observations of a single literary critic from 1977. Baumeister's opinion that 'there is little evidence that medievals engaged in introspection or experienced inner struggles'[70] can be refuted by considering the sorts of texts that were important to medieval audiences, including most spectacularly, Augustine's *Confessions*, which is entirely an account of the introspective struggle of coming to terms with knowledge, faith, memory and the self.[71] Augustine's ideas arose within a culture both temporally and geographically remote from that of medieval England, though, as a figure of enormous authority during the Western Middle Ages, his views were widely disseminated and influential. His dialogue concerning Natural Philosophy, his *Soliloquies*, was considered important enough to be one of the handful of books translated into Old English in the tenth century. The introduction to the vernacular version makes it quite clear that self-contemplation was very much on the agenda:

> Ða reahte he, hys mod for oft geasciende and smeagende mislicu and selcuð þing, and ealles swiðust ymbe hyne sylfne: hwæt he sylf wære, hwæþer hys mod and hys sawel deadlic were and gewitendlice.[72]
>
> Then he asked, his mind often querying and contemplating many strange things, most often concerning himself, what he himself was, and whether his mind and his soul were mortal and transitory.

Similarly, Boethius's *Consolation of Philosophy* (composed in 525) was translated into Old English during King Alfred's tenth-century educational reform, and explores the nature of the self. The vernacular translation retains, or perhaps elaborates, the focus on the self:

> Meaht þu nu ongitan hwæðer þu auht þe deorwyrðre hæbbe þonne þe selfne? . . . Ic wat, gif þu nu hæfde [*fulne anweald þines selfes, þon*] hæfde þu hwæthwugu on ðe self*um* þæs ðe þu næfre þinum willum alætan noldes, ne sio wyrd þe on geniman ne meahte.[73]

67 Danziger, 'The Historical Formation of Selves', p. 143.

68 Baumeister, 'The Self and Society', p. 193.

69 Baumeister, 'How the Self Became a Problem', p. 165.

70 Ibid.

71 For example, Book X reflects on memory, forgetfulness, mind and self: 'I find my own self hard to grasp' (X. xiv [24]), and 'In the vast hall of memory . . . I meet myself and recall what I am' (X.viii [14]). See St Augustine, *Confessions*, trans. Henry Chadwick (Oxford, 1992), pp. 193 and 187.

72 *King Alfred's Version of St. Augustine's Soliloquies*, ed. Thomas A. Carnicelli (Cambridge, MA, 1969), 1.1–3, pp. 48–9.

73 *King Alfred's Old English Version of Boethius, De Consolatione Philosophiae*, ed. John Sedgefield (Darmstadt, 1968), 9.2.19–24, p. 25.

> You might now consider whether you have anything more precious than yourself . . . I know, if you had full control of yourself, then you would have something always in yourself that you never willingly would relinquish.

Even more troublingly, eminent philosophers such as Hubert Dreyfus and Sean Dorrance Kelly claim in a very recent and widely read book that 'before Descartes people had little sense of an inner self', notwithstanding their earlier consideration of classical and Christian literature, including Augustine's *Confessions*.[74] This comment, like the generalised ideas on medieval selfhood enunciated by Baumeister and others, that there is a kind of standard collective pre-modern Western selfhood, can be effectively challenged not only on the grounds that the authors overgeneralise, but also through the provision of contrary evidence from literary texts produced and consumed within early cultures. Perhaps further perspective on modern views might be gained from reviewing historical variations on the self more rigorously, and by deploying emerging information, methodologies and ideas from cognitive approaches to literature and culture. This broadening conceptual and methodological basis might help to avoid the brief and dismissive summaries of medieval conceptions of the self offered in modern Psychology textbooks that make misleading assumptions.

Further, the poetic evidence presented below, pp. 147–59, shows how Literary Studies can contribute usefully to the Cognitive Sciences. Old English texts can show how deeply unconvincing entrenched ideas in Psychology can be. For instance, the offhand remarks that 'medieval Westerners did not wonder who they were or have identity crises the way modern ones do';[75] and 'in the religious imagery which was predominant at the time, an individual was not a separate identity, but was indissolubly linked to others, whose identity was collectively defined',[76] are, in the absence of supporting evidence, unsustainable.

Local constructions of the self can be recovered from what the writers of the Middle Ages have to say about selfhood, even if it is referred to under another name. In particular, secular rather than religious literary texts encoding cultural constructions of individual identity can provide information on pre-Modern self-imaging. Elegiac poetry is especially useful in this regard, as it provides information on inner identity through the expression of the thoughts and the feelings of the subject, well before the origins of the formal study of Psychology.[77] If the self is constructed through discourse, and 'past personal narratives' are significant in the creation of the self,[78] retrospective

[74] Hubert Dreyfuss and Sean Dorrance Kelly, *All Things Shining: Reading the Western Classics to Find Meaning in a Secular Age* (New York, 2011), p. 138.

[75] Roy F. Baumeister, 'The Self', *The Handbook of Social Psychology*, 4th edn, ed. Daniel T. Gilbert, Susan T. Fiske and Gardner Lindzey, 2 vols. (New York, 1998), vol. I, p. 726.

[76] Peter B. Smith and Michael Harris Bond, *Social Psychology across Cultures: Analysis and Perspectives* (New York, 1993), p. 95.

[77] As recognised by Susan M. Andersen and Lee Ross, 'Self-Knowledge and Social Inference: I. The Impact of Cognitive/Affective and Behavioral Data', *Journal of Personality and Social Psychology* 46.2 (1984), 280–93, at p. 280.

[78] Joseph P. Gone, Peggy J. Miller and Julian Rappaport, 'Conceptual Self as Normatively

literary narratives can provide important data on the imaginative representation of these practices and therefore on culturally enshrined views. As Elvin has shown with autobiographical poetry from ancient and medieval China, literary depictions of the self are valid counterpoints to contemporary views and selves from temporally remote cultures can be more recognisable than is imagined.[79] Specifically, literary texts can be interrogated for the medieval perspective on aspects of the self that have been assigned exclusively to modern cultures by psychologists: the self as the basis of moral authority; a hidden entity;[80] and a process. We can also perceive the early acknowledgment of the distinction between the ideal self and real self, and the self as 'observer and observed'.[81] It is obvious from a range of sources, including poetic ones, that medieval people did puzzle over the nature of the self and did formulate subjective experience in a way that we would find familiar.

Frequently, conceptualisations of the self are interpreted by modern scholars as being closely intertwined with local ideas about the mind, another culturally specific semantic construction.[82] Yet, although constructions of the self necessarily incorporate some element of mental awareness, these theories of subjective experience are generally believed to have been more heavily dependent upon the concepts 'mind' and 'soul' in pre-modern times than in contemporary Western cultures, especially in the understanding of the self as a 'reflective ego',[83] compared with our understanding of the self as consciousness and source of agency as well as the cognitive aspect of the person. One current theory, that the self is an enduring unchanging entity with executive control over the mind and body,[84] finds resonance in Old English poetry. In many pre-modern cultures, the constitution of personhood is perceived to be based on the relationship or interaction between mental functions and a sense of self. The relative value of the mind and the self, like the concepts themselves, are culturally specific. In medieval cultures, as in others, the mind and the self were constructed and interrelated in localised ways. Since cultural historians are still striving to understand the internal dynamics of ancient and medieval communities, we do not possess full knowledge of the social factors themselves that constitute 'schematising systems that affect and shape individual perception, memory, and thought'.[85] But because any viable model of subjectivity is self-perpetuating and participates in cultural transmission,[86] literary texts are rich sources of information

Oriented: The Suitability of Past Personal Narrative for the Study of Cultural Identity', *Culture and Psychology* 5.4 (1999), 371–98.

79 Elvin, 'Between the Earth and Heaven'.

80 See these and other categories in Baumeister, 'The Self and Society'.

81 Danziger, 'The Historical Formation of Selves', p. 147.

82 Wierzbicka, 'A Conceptual Basis for Cultural Psychology', p. 212.

83 Johnson, 'The Western Concept of the Self', p. 99.

84 Robin Harwood, *The Survival of the Self* (Aldershot, 1998), p. 7.

85 Jerome S. Bruner and Susan Weisser, 'The Invention of the Self: Autobiography and its Forms', *Literacy and Orality*, ed. David R. Olson and Nancy Torrance (Cambridge, 1991), pp. 129–47, at p. 134.

86 Recognised long ago by Dorothy Holland and Naomi Quinn, 'Culture and Cognition', in *Cultural Models in Language and Thought*, ed. Dorothy Holland and Naomi Quinn (Cambridge,

on a society's views on the self, even when this subject is not the ostensible topic of discussion.

Anglo-Saxonists who recognise this potential of literary texts as rich repositories of culturally specific information, stop short, however, of engaging with Cognitive Science in any thoroughgoing way. Instead, they treat 'the self' as a theme or a line of inquiry from a purely literary or cultural historical perspective.[87] A cognitive approach to these matters would ideally bring the two perspectives together, in order to enrich both disciplines. In order to demonstrate a way forward into this dialogue, the short elegiac lament poem, *The Wife's Lament* is examined from a Cognitive Literary/Cultural perspective for what it holds in respect of autobiographical memory and the self.

A cognitive approach to autobiographical memory and the self

There are many clear instances of textual representations of individual subjectivity in the literature of Anglo-Saxon England, and indeed many apparently conventional literary structures evident, whose rhetorical functions include the invitation to the reader to reflect on the fragmented or conflicted self as a subject of inquiry. These texts function in this way on both narrative and linguistic levels. For instance, many of the literary texts produced in Anglo-Saxon England feature a first-person speaker and a retrospective technique,[88] where the past is recalled to be compared with the present and the future, and the self is constructed by these discourses, an effective autobiographical memory function recognised by psychologists today.[89] Of course, the narratives are presented as fiction, not as true utterances, but they nevertheless must necessarily encode prevalent social views.[90] Within this narrative framework, the nature of personal responsibility for one's destiny is a dominant thematic preoccupation that entails discussion on free will, the soul, time, imagination and self-control. This last element, the expressed need for restraint, generally refers to a perceived bipartisan combative relationship between the self and the mind. The two are constructed as mutually dependent, controlling and potentially wayward constituents of the human person, a model of subjective existence which presupposes a primarily cognitive quality to the ego (as outlined by Ælfric), and constructs the self in a way

1987), pp. 3–40. See their cognitive approach to 'how cultural knowledge is organised' (p. 4).

87 Eugene Green, 'Speech Acts and the Question of Self in Alfred's *Soliloquies*', in *Interdigitations: Essays for Irmengard Rauch*, ed. Gerald F. Carr, Wayne Harbert and Lihua Zhang (Frankfurt am Main, Bern, and New York, 1998), pp. 211–18; Holly Elizabeth Jagger, 'Body, Text and Self in Old English Verse: A Study of "Beowulfian" and "Cynewulfian" Rhetoric', Ph.D. thesis, University of Toronto, 2002, DAI 63A (2003); Ronald J. Ganze, 'Conceptions of the Self in Augustine, King Alfred and Anglo-Saxon England', Ph.D. thesis, University of Oregon, 2004, DAI-A65/11(2005); Scott DeGregorio, 'Texts, *Topoi* and the Self: A Reading of Alfredian Spirituality', *Early Medieval Europe* 13 (2005), 79–96.

88 On retrospection, see Fred C. Robinson, 'Retrospection in Old English and Other Germanic Literatures', *The Grove* 8 (2001), 255–76.

89 Conway and Pleydell-Pearce, 'The Construction of Autobiographical Memories', p. 261.

90 Adamopoulos, 'Analysis of Interpersonal Structures', esp. p. 159.

which is in some respects analogous to modern constructions, though in some ways fundamentally different.

Poetic expression, as popular discourse, can be a viable form of case study,[91] though medieval literature has not been used in this manner by psychologists, and historians of Psychology likewise give Anglo-Saxon England short shrift.[92] Similarly, scholars of Old English literature who treat vernacular psychology keep very much within their field-specific boundaries, and pass up the opportunity of interdisciplinary inquiry. The recent book by Leslie Lockett, for example, *Anglo-Saxon Psychologies in the Vernacular and Latin Traditions*,[93] generally gives ideas and methods from contemporary Psychology a wide berth, notwithstanding its reference to 'Cognitive Linguistics and Transcultural Psychiatry',[94] its erudite engagement with earlier psychological traditions, and its close attention to vernacular concepts and language for the mind. Similarly, Malcolm Godden, who developed the earlier work of Peter Clemoes, has presented a convincing but monodisciplinary argument for two distinct psychological traditions operating in Anglo-Saxon England: one in the prose tradition, heavily dependent upon patristic and classical writers, which transmitted 'a unitary concept of the inner self, identifying the intellectual mind with the immortal soul and life-spirit'; and the second, in the vernacular verse tradition, which 'preserved the ancient distinction of soul and mind'.[95]

There is no doubt that these investigations are pertinent and scholarly, but they do leave the way open for integration with Cognitive Science, and in turn, can offer expertise in return. For example, the development of ideas about the self in relation to influential cultural texts and ideologies, and conflation with ideas about the soul could be introduced more thoroughly to Psychology. The Old English prose corpus provides some explicit discussions on the self, usually referred to as the soul (though of course not with precisely the same connotations as modern English *soul*), activated by human volition. Though the psychologist Baumeister describes the soul as 'an important precursor to the idea of an inner self',[96] there are stronger resemblances between the two than this remark suggests, resemblances that medievalists are equipped to pursue, especially given that semantic change over time misleads the current discussion about the self. The soul in Anglo-Saxon England is conceived of

91 Adamopoulos, 'Analysis of Interpersonal Structures'; Elvin, 'Between the Earth and Heaven'; Dorothy Holland, William Lachicotte Jr, Debra Skinner and Carole Cain, *Identity and Agency in Cultural Worlds* (Cambridge, MA, and London, 1998).

92 Most deal with the later Middle Ages from the time of Scholasticism: i.e. the twelfth century (Robert Pasnau, *Theories of Cognition in the Later Middle Ages* [Cambridge, 1997]); exclude England or Old English texts from their scope (Patrick J. Geary, *Phantoms of Remembrance: Memory and Oblivion at the End of the First Millennium* [Princeton, 1994]; and Simon Kemp, *Medieval Psychology* [New York, 1990]); or, more problematically, compare earlier societies with stages of infant cognitive development (Charles M. Radding, *A World Made by Men: Cognition and Society, 400–1200* [Chapel Hill and London, 1985]).

93 Leslie Lockett, *Anglo-Saxon Psychologies in the Vernacular and Latin Traditions* (Toronto, 2011).

94 Ibid., p. 9.

95 Malcolm Godden, 'Anglo-Saxons on the Mind', in *Learning and Literature in Anglo-Saxon England*, ed. Michael Lapidge and Helmut Gneuss (Cambridge, 1985), pp. 271–98, at p. 271.

96 Baumeister,'The Self and Society', p. 195.

as far more than an immortal spirit, and at times encompasses many of the qualities usually associated with the self in Cultural Psychology, a disciplinary dynamic that could be developed further. The evidence marshalled by Godden, for example, suggests that conception of the self changed once Christianity and Latin literacy were introduced into England, that even within the period in which England was 'Anglo-Saxon', there is a discernible historical development or change in the conception of the self: the once separate psychological source of agency and the cognitive apparatus of the poetic tradition were melded into a single entity within the prose tradition.

For instance, the prolific homilist, Ælfric of Eynsham enunciates the 'combined' view of the Anglo-Saxon period, describing the functions of the soul as both cognitive and volitional. Frequently in the context of matching Latin terms with suitable Old English equivalents or situating the soul as the complement to the body, Ælfric conceptualises the human soul in the context of Christian lore:

> Seo sawul hæfð . . . on hire gecynde þære halgan þrynnysse anlicnysse, on þan þe heo hæfð gemynd, and andgit, and wyllan. An sawul is, and an lif, and an edwist.[97]
>
>> The soul has . . . in its nature a likeness to the Holy Trinity, in that it has memory, and understanding, and will. It is one soul, and one life, and one substance.

He explicitly calls the soul a 'gesceadwis gast' (line 171), 'rational spirit', which 'governs the five senses as well as the body (lines 195–6). We can see at work the task of pairing Latin and Old English words, when Ælfric refers to the various Latin terms for the soul depending upon which faculty is being invoked (lines 181–9):

> Hyre nama is anima þæt is sawul and seo nama gelympð to hire life. And spiritus gast, belimpð to hire ymbwlatunge. Heo is sensus, þæt is andgit, oððe felnyss, þonne heo gefret. Heo is animus, þæt is mod, þonne heo wat. Heo is mens. þæt is mod, þonne heo understent. Heo is memoria, þæt is gemynd, þonne heo gemanð. Heo is ratio, þæt is gescead, þonne heo tosceat. Heo is uoluntas, þæt is wylla, þonne heo hwæt wyle. Ac swa þeah ealle þas naman syndon sawul.
>
>> Its name is *anima,* that is, Soul, and the name befits its life; and *spiritus,* that is spirit, which pertains to its contemplation. It is *sensus,* that is perception or sensation, when it perceives. It is *animus,* that is, intellect, when it knows. It is *mens,* that is, mind, when it understands. It is *memoria,* that is, memory, when it remembers: It is *ratio,* that is, reason, when it reasons. It is *voluntas,* that is will, when it wills anything; nevertheless all these names are one soul.

Typically of the prose writers who develop their ideas from Christian Latin texts, Ælfric acknowledges the complex functions of the soul and insists that the faculties are elements of one guiding essence. There is no reference to the

97 Ælfric, 'The Nativity of Our Lord Jesus Christ', ed. Walter W. Skeat, *Lives of the Saint*, 2 vols (1844; repr. Woodbridge, 1999), vol. 1, pp. 10–25, at p. 16, lines 112–15.

self as a multiple, variable entity such as we find in modern discussions,[98] but rather to a multivalent entity with variable attributes. As Danziger notes, in Christian, medieval thought, the soul's 'indivisibility was of the essence',[99] but this comment should not imply that its function was perceived of as singular, a distinction that medievalists can bring to the attention of psychologists.

The earlier prose writers, Alcuin of York and Alfred, king of Wessex, likewise indicate the inadequacy of evaluating medieval philosophical or psychological ideas based on semantic equivalence with modern-day terms. Alcuin equates the soul with the intellect, the rational mind, and Alfred uses the words *sawl*, 'soul', and *mod*, 'mind', interchangeably, though he and other Anglo-Saxon writers demonstrate that the words are by no means always synonymous.[100] Clearly, a correspondence between reason and the inner self was perceived in this society, and the semantic nuances of terms for these concepts, as well as 'soul', are very context specific. The writings of Alcuin and Alfred, as well as Ælfric, demonstrate that philosophical inquiry into the self was alive and well in Anglo-Saxon England.

Furthermore, the self was a feature of everyday discourse in Anglo-Saxon England, an idea that can be demonstrated through the tools of historicised linguistic study. Even though the word *self* was rarely used as a noun in this culture (but see *The Wife's Lament*, below, pp. 148–51), it is there frequently as an adjective and pronoun, and as an element in several compound terms: e.g. *selfbana*, 'a suicide' (also *selfcwalu*, 'self-slaughter; and *selfmyrthere*, 'a suicide'); *selfdema*, 'one who depends on his own judgement'; *selflic*, 'of one's own accord'; *selflice*, 'selflove' (arrogance which can 'grip the mind');[101] *selfwill*, 'self-will, free will'; *selfwille* and its variations 'voluntary' (also *selfwendlice*, 'under one's own direction'; and *selfwildlice*, 'under one's own control').[102] These linguistic facts, particularly the existence of several terms for concepts such as 'self-judgement' and 'free will', remove any doubt that a sense of self as a guiding entity was conceptualised and utilised, at least in the learned scribal culture of Anglo-Saxon England which has provided most of our extant documents (specifically in the legalistic contexts in which these terms are often found). Indeed, the existence of compounds based on *self* in modern English usage

98 For example, Seymour Rosenberg, 'Multiplicity of Selves', in *Self and Identity: Fundamental Issues*, ed. Richard D. Ashmore and Lee Jussim (New York and Oxford, 1997), pp. 23–45.

99 Danziger, 'The Historical Formation of Selves', p. 147.

100 See Michael Joseph Phillips, 'Heart, Mind, and Soul in Old English: A Semantic Study', Ph.D. thesis, University of Illinois, 1985, DAI-A 46/07 (1986); and Antonina Harbus, *The Life of the Mind in Old English Poetry* (Amsterdam and New York, 2002), pp. 34–7.

101 The expression is a translation of the Latin 'amor proprius cum rectoris mentem ceperit' in the Old English translation of Gregory's *Pastoral Care* , see BT, BTS, s.v. *selflice*, and elsewhere is a gloss to *superbia*, 'arrogance'. Cf. the description of Adam in *Genesis B* as a 'selfsceafte guma' (line 523) a 'self-determined man'.

102 BT, BTS, s.vv. Unlike modern English, which includes hundreds of compounds based on *self*, in Old English, there were relatively few, and most are related to personal direction (though there are a few exceptions: e.g. *selfæta*, 'a self-eater, cannibal'; *selfæte*, 'wild oat'; *selfe*, 'in the same way'). That there are three terms relating to suicide reflects a legalistic as well as moral identification of the object of personal violence. Similarly the existence of a noun, two adjectives and two adverbs connoting 'self-will' suggests the importance of this concept in this culture.

has been interpreted as evidence of the perception of an entity other than the soul,[103] though this idea can be applied just as appropriately to earlier cultures, despite its Anglocentrism. That *self-* compounds and 'mind' and 'soul' words are used widely throughout the existing Old English corpus of texts suggests that Anglo-Saxon writers and readers were interested in the self as a subject of inquiry, including in the imaginative output of poetry. This context makes the self even more prominent, because when literary considerations come into play, the linguistic freight of all these terms is even more important.

The imaginative discourse of poetry provides more fruitful and less consciously soul-based ideas on the self and its relation to the mind, especially since, unlike the later prose texts of Anglo-Saxon England, much of the verse is thought to have existed in some oral context prior to the importation of Latin literacy and is therefore capable of transmitting secular as well as Christian views. Because the dating of Old English material is extremely problematic and we possess only the written forms of the texts, it is impossible to distinguish earlier strands of influence, so scholarly practice heeds the empirical evidence of manuscript dating and context more than conjecture concerning pre-literate forms of narratives.

There is also a terminological consideration. As the quotation from Ælfric demonstrates above, *self* is not used in most of the modern senses, and the proliferation of vernacular terms for the mind and its faculties (e.g. *mod, sefa, modsefa, ferþ, gemynd*) and the soul (*sawl, gast*) makes it very difficult to identify when the purely cognitive faculty is being distinguished from the essential entity of a person or source of agency, and even when the two are conceived of as identical to one another. Most often, the self is expressed with personal pronouns, 'he', she' and 'I', but occasionally the 'mind' and 'soul' terms are used in contexts which seem to imply a connotation of 'self'. Because a remote foreign language is involved, the data is very difficult to interpret as it is in cross-cultural studies today.[104]

Memory and self-perception in The Wife's Lament

The Wife's Lament, as a narrative account of one woman's personal circumstances, presents a clear insight into Anglo-Saxon perceptions of the self. The first person speaker opens her narrative with the following lines which equate experience and feelings with the self: 'Ic þis giedd wrece bi me ful geomorre, / minre sylfre sið' (1–2a), 'I will narrate this tale about very sad me, the experience of my self'. This citation includes a very rare use of the word *self* as a substantive, and emphatically communicates a sense of self as an enduring and individual entity shaped by experience. This introduction calls attention to the subjective mode of discourse and objectifies the self through the explicit reference to personal history as narrative. The evidence of similar openings to other poems suggests that the Anglo-Saxons had a

103 Danziger, 'The Historical Formation of Selves', p. 154.

104 For example, Markus and Kitayama, 'Culture and the Self'.

taste for lament poetry within an autobiographical format. *The Seafarer* opens with an unequivocal announcement that also employs a *me sylf* construction: 'Mæg ic be me sylfum soðgied wrecan, / siþas secgan' (1–2a), 'I will sing a true song about myself, speak of my experiences'; and *Resignation B* is also explicitly autobiographical: 'Ic bi me tylgust / secge þis sarspel ond ymb siþ spræce' (96b–97), 'I narrate this sad story entirely about myself and speak about a journey, eager with longing, and think about the sea'. These remarks both catch the attention at the outset and also communicate that the tale will be about the speaker, narrative and usually mournful. The use of the first-person pronoun, 'I', also sets up a particular model of truthful personal expression which characterises much of the extant Old English elegy, and uses a standard way of marking expressions as personal utterances focused on the self. Here, the 'I', the self, is distinct from the soul, the body and the thinking mind, though it acts in close concert with them in the folk conceptualisation of the person.

In *The Wife's Lament*, the reason for the woman's grief is her separation from her lover, though she is also isolated from her usual social group, alone in a grove. After recounting her hardships and solitary life in a cave, she compares her lot with that of lovers who are permitted to remain together (32b–41):

Ful oft mec her wraþe begeat
fromsiþ frean. Frynd sind on eorþan,
leofe lifgende, leger weardiað,
þonne ic on uhtan ana gonge
under actreo geond þas eorðscrafu.
Þær ic sittan mot sumorlangne dæg,
þær ic wepan mæg mine wræcsiþas,
earfoþa fela; forþon ic æfre ne mæg
þære modceare minre gerestan,
ne ealles þæs longaþes þe mec on þissum life begeat.

Very often the departure of my lord has cruelly taken hold of me here. There are friends on earth, dear ones living, occupying a bed, while I walk alone in the dawn under the oak-tree throughout these caves. There I can sit for a summer-long day, there I can weep for my miseries, for many hardships. Therefore I can never rest from my anxieties of mind, nor from all this longing which this life has begotten for me.

Here is represented the reflective account of a woman who is not only experiencing inescapable inner struggles, but calmly acknowledging them and their effect upon her psyche. These lines imply an independent construction of the self, which is perceived to be separate from others and created by personal experience,[105] not a sense of collective identity[106] understood by some psychologists and historians to have existed in 'less complex' earlier societies.[107] The speaker's emotional balance seems to have been upset by the

105 Markus and Kitayama, 'Culture and the Self', p. 224.
106 As argued by Baumeister, 'How the Self Became a Problem', esp. p.165.
107 Triandis, 'The Self and Social Behaviour', p. 509.

loss of a key relationship, a recognisable feature of lament poetry in general. She also admits the necessity of concealing inner reality from the world and appearing outwardly more composed than she feels (a recurrent them throughout Old English elegies) when she imagines the mental dilemmas of her lover in an act of projected mind-reading (42–53):

> A sceal geong mon wesan geomormod,
> heard heortan geþoht, swylce habban sceal
> bliþe gebæro, eac þon breostceare,
> sinsorgna gedreag, sy æt him sylfum gelong
> eal his worulde wyn, sy ful wide fah
> feorres folclondes, þæt min freond siteð
> under stanhliþe storme behrimed,
> wine werigmod, wætre beflowen
> on dreorsele. Dreogeð se min wine
> micle modceare; he gemon to oft
> wynlicran wic. Wa bið þam þe sceal
> of langoþe leofes abidan.

> Ever must the young man be sorrowful, and the thought of his heart harsh. Likewise, he must have a cheerful demeanour, along with the care of the heart, a constant multitude of sorrows, (and) may all his worldly joy be dependent on him alone, may he be outlawed widely in a distant tribal land, since my friend sits under a stony cliff, covered with hoar-frost by the storm, my heart-weary lord, surrounded by water in a dreary hole. He, my lord, experiences great sorrow of mind. He remembers too often a more joyful dwelling. Woe it is for the one who must await the dear one in longing.

This passage can be interpreted in several different ways, so this translation is necessarily speculative, but however the agency and identity of the people being referred to are construed, it is clear that the speaker has a distinct conception of the self as source of individual identity – a self-reliant centre of consciousness and agency, able to be at least partially concealed from the world. The female narrator also assumes the individual has the ability to imagine the mental organisation and emotional reactions of other individuals to read minds and to impute mental states to others. Her acknowledgment of the youth as being 'dependent upon himself for all worldly joy' further indicates the conception of the self as a private entity, emotionally independent. Like Elvin's medieval Chinese poetry, this Old English poem presents a coherent and recognisable view on the self that resonates with modern self constructions, and shares many of its attributes, including self-knowledge and an awareness of psychological processes, as delineated by Baumeister. Emotion is embodied, symbolised by surroundings, able to be imagined by other minds, and represented in allusive verse. More specifically, an over-riding mood (here, woe, or sorrow) is imagined to arise form personal circumstances and to permeate lived experience, and emotions give rise to mental functions and perhaps by extension to inclination and character attributes: 'ever must the thoughts of his heart be harsh'.

Danziger, who ascribes the origins of the Western idea of the self to 1300, claims that what he perceives as the 'opposition between the self as an incarnation of wickedness and divine goodness survives for several centuries',[108] and that it was not until the early seventeenth century that 'the self is conceptualised as a self-reflective monitoring agent oriented to the maximising of its own overall advantage'.[109] There was, he argues, a perceived need for the self to exercise some sort of control over the mind in order to achieve the desirable spiritual goal as well as mental tranquillity, and also an awareness that this action was occurring. But *The Wife's Lament*, written down in the tenth century and probably composed and recited orally much earlier than that, presents compelling evidence against Danziger's dating of the origins of ideas about the self to around 1300. In this text, as in several other elegiac Old English poems, one has the personal responsibility of restraining the mind in its receptacle (the chest, as it was perceived in this culture), not allowing imagination to wander too freely. There are some similarities here with the Newar people studied by Parish, who consider the mind to be situated in the heart (the locus of cognitive as well as emotional experience), and construct a model of self which can both experience and have a degree of agency over mental functions.[110] Moreover, the model of the self embedded in the narrative of *The Wife's Lament* extends the influence of that agency beyond the present apprehension of memories, current self-control, into an imagined future narrative for herself and others. This future organisational perspective accords with contemporary findings about autobiographical memory and an imagined future self.[111]

Discursive and imagined selves in Old English poetry

In Old English, the self is rarely encoded in a specific single word; usually the expressed subject, 'I', represents the essential entity of the human being as it does in later discourse.[112] For example, in *The Wanderer* and *The Seafarer*, discussed above, the self is positioned outside the mind and has consciousness and volition but is in no sense 'an incarnation of wickedness'. That the self is an independent entity, is comprised of individual experience, and can be apprehended only by a similar self-construction, is implied in *The Wanderer* when the speaker says he is seeking 'þone þe in meoduhealle   min mine wisse' (27), 'the one who might recognise my mind in the meadhall'.

108 Danziger, 'The Historical Formation of Selves', p. 143.

109 Ibid.

110 Parish, 'The Sacred Mind', p. 345.

111 T. Suddendorf and M. C. Corballis, 'The Evolution of Foresight: What is Mental Time Travel, and is it Unique to Humans?', *Behavioral and Brain Sciences* 30.3 (2007), 299–351; and Clare Rathbone, Martin A. Conway and Chris J.A. Moulin, 'Remembering and Imagining: The Role of the Self', *Consciousness and Cognition* 20.4 (2011), 1175–82.

112 Triandis, 'The Self and Social Behaviour', p. 506; and Rom Harré and Grant Gillett, *The Discursive Mind* (Thousand Oaks, London and New Delhi, 1994), p. 106.

Similarly, in *The Seafarer*, there is a disjunction between the knowledge of the land-dweller and the Seafarer, which causes an insurmountable epistemological problem (12b–15):

> Þat se mon ne wat
> þe him on foldan fægrost limpeð,
> hu ic earmcearig iscealdne sæ
> winter wunade wræccan lastum.

> The man who lives on the land most happily does not know how I, full of sorrows, remained on the ice-cold sea during winter in the paths of exile.

This perceptive disjunction becomes a major motif of the Seafarer's lament, and is repeated at lines 27–30 and 55b–57. Again, the self is construed as sad rather than sinful, and self-knowledge is at the rhetorical and aesthetic centre of this text.

In *The Wanderer* above, the verbal reiteration (and semantic variation in the original) on 'mind', the object of speech, venue of emotions and volatile faculty requiring energetic restraint by the self, places individual identity in the psychological context. The implied dislocation of self and mind in this poem allows the self to situate its subjectivity outside the mind and acknowledge the desirability of self-control. The mind can be directed by the self in this schema, the ultimate moral authority: A person must 'hold his mind, think what he will'. This poetic expression of the desirability of reticence, like the sentiments of *The Wife's Lament*, encodes the construction of an independent self perceptible in Western cultures today,[113] though the proverbial expression of normative behaviour 'righteous ones must often . . .' shows the enduring force of ingroup goals,[114] even when the individual is remote from the usual group. The poetic narrative reinforces those desirable conventions through the fictive construct of a solitary individual, similar to the female narrator of *The Wife's Lament*. Similarly, the point of view of the speaker, in observing his own mind and decision-making processes, demonstrates a degree of self-awareness. In both poems, we can see the poetic exploration of an individual away from the usual social setting, and the 'discursive' construction of the self 'through the private discourse of retrospective personal narrative'.[115]

The self has priority over the mind in this schema. *The Wanderer* enunciates the relationship that construes the self as the dominant cognitive apparatus (70–2):

> Beorn sceal gebidan, þonne he beot spriceð,
> oþþæt collenferð cunne gearwe
> hwider hreþra gehygd hweorfan wille.

113 Markus and Kitayama, 'Culture and the Self', p. 224.
114 Triandis, 'The Self and Social Behaviour', p. 509.
115 Harré and Gillett, *The Discursive Mind*, p. 104.

> A man must wait, when he utters a boast, until the brave one knows for certain where the thoughts of his heart will turn.

The self is constructed as being only partially aware of the motives behind the mind's state (58–60):

> Forþon ic geþencan ne mæg   geond þas woruld
> for hwan modsefa   min ne gesweorce,
> þonne ic eorla lif   eal geondþence.

> Therefore, I cannot think for all the world why my mind does not darken when I fully consider the life of men.

These lines express a distinction between the conscious and unconscious self,[116] and the mind and the self as two thinking entities, a disjunction that is unfamiliar to us today. This model of the self possessing executive control over the mind appears to have been conventional: it occurs in *The Seafarer*: 'Stieran mon sceal strongum mode,   ond þæt on staþelum healdan' (109), 'A man must govern an unruly mind and keep it in (its) place', and is repeated in almost the same words in an Old English gnomic text known as *Maxims I*.

A similar model of the self is implied in the poem *Vainglory* (9–11):

> Þæt mæg æghwylc mon   eaþe geþencan,
> se þe hine ne læteð   on þas lænan tid
> amyrran his gemyndum   modes gælasan.

> Each man is able to consider this easily, who does not allow his mind's worldly care in this transitory period to hinder him in his purpose.

These lines contradict the view that prior to the seventeenth century, the soul, not the self, was the guiding entity of the individual, as argued by Danziger and others. Rather, in the folk model of the person articulated in these lines, the self holds ultimate responsibility for action, thought and for 'maximising his own overall advantage'.[117]

Like our contemporaries, the medieval writers responsible for these laments present characters who are aware of their motives for behaviour within an independent conception of the self. A great deal of this poetry centres on solitary individuals rather than those within a group and focuses on the private self rather than the public or collective self.[118] The elegiac thoughts of the person separated from their usual ingroup tend to focus on loss and misery, construing the separation as a negative turn of events. Autonomy appears to have been thrust upon these poetic personae rather than chosen by them, and this situation is the core of the 'lament' experience of this poetry which recounts an 'evolving self' adapting as a result of changing circumstances,[119] though with the unfamiliar risk of the self being in conflict with the mind. So,

116 Godden, 'Anglo-Saxons on the Mind'.
117 Danziger, 'The Historical Formation of Selves', p. 143.
118 Triandis, 'The Self and Social Behaviour', p. 506.
119 Gone *et al.*, 'Conceptual Self as Normatively Oriented', p. 384.

rather than construe medieval people as possessing stable attributes,[120] this evidence encourages us to believe that these selves were as capable of change and of observing that change as we are today. Of course, social institutions might have been organised differently, but it would be a mistake to confuse this phenomenon with the construction of medieval identities.[121]

In several Old English elegiac narratives, a speaker laments his or her separation from the group. The narrator of *The Wife's Lament* grieves for her lover and their usual social context, and the speaker of *The Seafarer* mourns his alienation at sea. In *The Wanderer*, the ritual gift-giving ceremony in the hall is recalled, along with the speaker's affection for his lord. The remembered self is 'sociocentric',[122] in that social norms and responsibilities guide behaviour, though this is contrasted with the present circumstances of the solitary and self-oriented narrator. The view that the self shapes emotional experience here is independent according to the formulation of Markus and Kitayama: the emotions, especially grief, are ego-focused, and other selves are perceived to be outside the bounds of the self rather than connected to it.[123] The poems frequently depict this personal isolation by means of the metaphor of physical isolation, most strikingly in *Wulf and Eadwacer*: 'Ungelic is us. / Wulf is on iege, ic on oþerre' (3–4), 'A difference exists between us. Wulf is on one island, I on another'.

Old English elegies also provide evidence of the existence of the idea of the self as a hidden entity, common in later discourses. The belief that only in early-modern times might a person be aware of presenting a different self to the world from the one kept hidden[124] is not supported by the Anglo-Saxon evidence. Although the word 'mind' is used in the translations, the psychological entity referred to in the Old English shades in meaning towards our 'self'. In *The Wife's Lament*, a would-be murderer keeps his mind hidden: the woman's lover is described as 'concealing his mind, considering murder' (20). Similarly, the possibility of self-concealment and the desirability of mutual frankness is encoded in the following lines of the gnomic poem, *Maxims I* (1b–3):

> Ne læt þinne ferð onhælne,
> degol þæt þu deopost cunne! Nelle ic þe min dyrne gesecgan,
> gif þu me þinne hygecræft hylest ond þine heortan geþohtas.

> Do not let your mind be concealed, that which you know most deeply to be unknown. I will not speak my mystery if you hide your reason and the thoughts of your heart from me.

The above quotations from *The Wanderer* and *The Wife's Lament* demonstrate that the self is conceived of as an internal entity, hidden from the outside

120 Baumeister, 'How the Self Became a Problem', p. 164.
121 As argued by Spiro, 'Is the Western Conception of the Self "Peculiar"'.
122 Tory E. Higgins, 'The "Self-Digest": Self-Knowledge Serving Self-Regulatory Functions', *Journal of Personality and Social Psychology* 71.6 (1996), 1062–83, at p. 1077.
123 Markus and Kitayama, 'Culture and the Self', p. 235.
124 Baumeister, 'The Self and Society', p. 196.

world, and that emotions are restrained by the mind. Other expressions within the corpus of the need for keeping one's own counsel and restraining gregarious tendencies confirm the desirability of private reflection as a cultural and literary trope. For example, in a poem entitled *Homiletic Fragment II*, the speaker says: 'heald hordlocan, hyge fæste bind / mid modsefan' (3–4a), 'Hold the treasure-chamber (mind), bind fast your thought within the mind'. The warning is repeated in *The Fortunes of Men*, though here the mind is the faculty of control: 'þonne he gemet ne con / gemearcian his muþe mode sine' (52b–3), 'Then he does not know moderation, to restrain his mouth with his mind'. Again, two levels of consciousness and control are perceived, and the necessity of imposed restraint enunciated. These expressions suggest a perception of human identity situated upon the combative relationship between natural or unconscious thought and a higher level of controlled or willed thought, an aggressive variation on our perception of self-control.[125] They also express social norms and the desirability of conforming to them, as well as the awareness of a disjunction between the inner and outer selves.[126] The inner self is private and not to be put on display.

Old English poetry provides evidence of some interest in the dynamic and process-oriented view of the self that is common in modern Psychology. In the earliest English model of the mind, the onus is on the individual to rule the mind collaboratively with the self in order to direct thoughts appropriately. These texts construct a picture of the self that ideally acknowledges the desire for change, and sets about achieving the program of reform imagined. The poems dwell on the moods and attitudes of their first-person speakers and construct narratives based on memory and contemplative self-reform.

These texts express an awareness of the motivation to be in full control of the agency of ones actions[127] and reflect the tendency of self-awareness to be directed at regulating the self.[128] They are concerned with the paradox of a person's striving to be better than him- or herself, or at least trying to improve the interpretation of present circumstances, phenomena observable in later cultures including our contemporary Western ones.[129] The idealised self is the one constant within the variables of time, place and multiple selves, but it can be guided and developed if only the mind can be controlled. This inconvenient waywardness of mind causes a tense symbiotic power relationship of human definition. The texts encode the development as the mind alters in response to changing circumstances and the self becomes another. The elegies enact the psychology of consolation, and the past is not so much

125 As argued by C. S. Carver and M. F. Scheier, *Attention and Self-Regulation: A Control Theory Approach to Human Behavior* (New York, 1981); and Howard Rachlin, *The Psychology and Economics of Self Control* (Cambridge, MA, 2000), who argues for a model of reinforcement to bring about behavioural change.

126 Baumeister, 'How the Self Became a Problem': 'self equated with social, public self' (p. 164).

127 Markus and Kitayama, 'Culture and the Self'.

128 As recognised by Higgins, 'The "Self-Digest"', 'self-knowledge serves self-regulatory functions', at p. 1062.

129 As argued, for instance, by M. R. Banaji and D. A. Prentice, 'The Self in Social Contexts', *Annual Review of Psychology* 45, ed. L. Porter and M. Rosenzweig (1994), 297–332.

visited or preserved as used as a catalyst to change the present, and therefore, the self on view is a form of process. The apparent desired outcome of this process is the refinement of the present self and the regulation of the mind, as has been observed in later cultures.[130]

Besides self-awareness and consciousness of the engagement of cognitive functioning in the self-process, retrospection is evident in Old English elegiac poetry. It is through stepping out of the narrative present into the recalled past that a process of acknowledgment and change is initiated in these narratives. The use of verbal tense distinctions to enunciate a temporal development of the self is one mechanism whereby this is encoded. As Seltzer describes the sequence of *The Wanderer*: 'The quest that began in the past . . . through the use of memory, becomes fully cerebral in the predominantly present-tense analysis.'[131] All the elegies contain narratives structured on reminiscence: the entire textual story is the product of memory, analysed in the present. The use of the past tense for some passages of the narratives, juxtaposed to the change to the present tense, immediately indicates not only a retrospective, but also a contemporary analysis of the past. The narrator in this way fashions a narrative of a former self and sets up a model of the temporal development of the self, that is, the deliberate enactment of spiritual change over time with individualistic goals rather than collective ones. This disjunction or at least an alienation from the self is made explicit at the outset of two of the elegies, *The Wife's Lament* and *The Seafarer* mentioned above. In these cases, the past has been fashioned into a tale, a narrative that can be contemplated by another person. Likewise, the speaker of *Deor* introduces the conclusion to this poem with a signal of autobiographical truth and self-conscious observation: 'þæt ic bi me sylfum   secgan wille' (35), 'I will say this about myself', after the communication of biographical details and the narrative use of the third-person: 'on sefan sweorceð,   sylfum þinceð / þæt sy endeleas   earfoða dæl' (29–30), 'his mind clouds over; it seems to him himself that his share of sorrows will be endless'. The concluding refrain of this poem, which also ends each stanza is the observation that 'þæs overeode,   þisses swa mæg!' (42), 'That situation passed away; so may this one'. This line suggests that the self, the centre of consciousness, has been renegotiated to accommodate the new circumstances, an idea very familiar to dwellers in contemporary Western societies.[132]

Within this schema of development is the opposition of the real self with an imagined ideal of selfhood, a basic premise of the Christian tradition of confession, recognised today as a dominant medieval 'technology of the

130 See Harwood, *The Survival of the Self*; and Conway and Pleydell-Pearce, 'The Construction of Autobiographical Memories'.

131 John L. Seltzer, '*The Wanderer* and the Meditative Tradition', *Studies in Philology* 80 (1983), 227–37, at p. 237.

132 'Selfing is inherently a unifying, integrative, synthesizing process', see D. P. McAdams, 'The Case for Unity in the (Post)Modern Self: A Modest Proposal', in *Self and Identity: Fundamental Issues*, ed. Richard D. Ashmore and Lee Jussim (New York and Oxford, 1997), pp. 46–78, at p. 56.

self',[133] and perhaps a precursor to modern Psychotherapy.[134] The acknowledgment of the self of the confession – as a moral agent with the responsibility of self-monitoring and self-improvement – is present in the autobiographical narratives of some Old English poetry. The speaker of the poem *Resignation B* enunciates his desire to trust himself and for the self to make preparations for the soul (70–8a):

> Hwæþre ic me ealles þæs   ellen wylle
> habban ond hlyhhan   ond me hyhtan to,
> frætwian mec on ferðweg   ond fundian
> sylf to þam siþe   þe ic asettan sceal,
> gæst gearwian,   ond me þæt eal for gode þolian
> bliþe mode,   nu ic gebunden eom
> fæste in minum ferþe.   Huru me frea witeð
> sume þara synna   þe ic me sylf ne conn
> ongietan gleawlice.

> Yet in all this I wish to have courage and to rejoice and hope for myself, adorn myself for the spiritual path and set myself to that journey which I must undertake, prepare my soul and endure all this myself for God with a happy mind, now that I am firmly resolved in my heart. Indeed the Lord knows of certain sins of mine that I do not know to perceive wisely myself.

The references to 'myself' in this quotation, as well as the expression of a wish for an ideal mental and spiritual state, set the self apart as a separate, directing entity with incomplete knowledge of transgression in a traditional Christian model of spiritual life. This articulation of the moral consciousness situates the ethical care of the self as the chief concern of the thinking person. The grouping of the words 'self', 'spirit', 'mind', as well as the pronoun 'I' within these lines shows that the concept of the self, embracing the cognitive, emotional and spiritual aspects of human identity, is a matter of emphatic concern to this poet. Because we have here a different model of the self from those that dominate our contemporary discussions, the linguistic barrier to interpretation becomes even more problematic, but the poetry quoted above confirms the existence of a culturally specific self schema in Anglo-Saxon England which acknowledged an individualistic inner self rather than a more 'collective' self (as psychologists like Baumeister postulate). Personal responsibility is integral to this model of subjective experience and this involves self-awareness, self-maintenance and the development of an ideal self through restraint and constant renegotiation.

The quotation directly above sets up a 'bipolar' self as subjective and

[133] Danziger, 'The Historical Formation of Selves', p. 150, using concepts from Foucault, *Technologies of the Self* (Boston, 1988).

[134] John D. Richardson and Destin N. Stewart, 'Medieval Confession Practices and the Emergence of Modern Psychotherapy', *Mental Health, Religion and Culture* 12.5 (2009), 473–84. For a different, cross-disciplinary perspective on the medieval legacy, see also Noel Packard and Christopher Chen, 'From Medieval Mnemonics to a Social Construction of Memory: Thoughts on Some Early European Conceptualizations of Memory, Morality, and Consciousness', *American Behavioral Scientist* 48.10 (2005), 1297–319.

objective entity, both a 'passive experiencer' and an 'agent of psychological states'.[135] The textual situation of subjectivity, outside both the self and the mind, and encoded in the use of first-person narration, shows that the 'now' self exists outside these entities and also beyond the past selves visited in the narrative. The changing points of view articulated in *The Wanderer* dramatise this self-observation and demonstrate that absolute individual isolation is a prerequisite for contemplating past and possible future selves. After a first-person monologue, the narrator says: 'swa cwæð snottor on mode, gesæt him sundor æt rune' (111), 'Thus spoke the one wise in his mind (or 'thus spoke the wise one in his mind'); he sat himself apart in private meditation'.

In other contexts, the self is observing itself in past incarnations. In this objectification of the self, mental pain is expressed as physical torment through the reiterative use of this highly charged psychological vocabulary, especially within metaphorical structures encoding binding or restriction. The speaker of *The Riming Poem* says (43–50):

> Nu min hreþer is hreoh, heofsiþum sceoh,
> nydbysgum neah; gewiteð nihtes in fleah
> se ær in dæge wæs dyre. Scriþeð ne deop in feore
> brondhord geblowen, breostum in forgrowen,
> flyhtum toflowen. Flah is geblowen
> miclum in gemynde; modes gecynde
> greteð ungrynde grorn efenpynde,
> bealofus byrneð, bittre toyrneð.

> Now my mind is troubled, timid in times of afflictions, near to distress. That which in the day was excellent departs now in flight. Now deep within his soul moves an advancing secret fire, growing in his breast, widely spread. Wickedness has grown greatly in mind; a bottomless, unrestrained sorrow takes hold of the nature of the mind, it burns terribly, grows savagely.

The speaker does not say 'I am troubled' but rather 'my mind is troubled', distinguishing the mind from the self. The metaphor of sorrow as a consuming fire suggests the individual's absolute vulnerability to the fragile psychological world that can be assailed by external influences: the self as passive experiencer. The externalisation of this torment exonerates the mind from blame for its own condition. Similarly emotions are 'taken on' rather than passively experienced,[136] and are apprehended actively.[137] But the self is also an active moral agent in that it is acknowledging the impact of emotions upon itself and shaping reality by controlling the mind and by enunciating that recognition in a poetic narrative. The self remains the centre of consciousness and basis of identity, but relinquishes to stronger forces its control as a source of agency, a chief difference between secular

135 Parish, 'The Sacred Mind', pp. 340–1.

136 Godden, 'Anglo-Saxons on the Mind', p. 286.

137 Peter Clemoes, *Interactions of Thought and Language in Old English Poetry* (Cambridge, 1995), p. 365.

and Christian constructions of the self.

Frequently, the elegies express sadness or distress, and invite an empathetic response. Because ideas about the self and the affective life are so inter-related, an understanding of how emotions were conceived by the Anglo-Saxons will shed some light on their views on the self. Nicholson has examined the emotional vocabulary of Old English poetry and prose and has concluded that the Anglo-Saxons 'had some appreciation of how distress affected mind and body', and in particular, that 'Old English distinguished between what would now be regarded as psychological and somatic expressions of emotional distress'.[138] Elegiac poems like *The Wanderer* and *The Seafarer* not only communicate an awareness of this difference, but rely on somatic (physical) metaphors based on heat, cold and restriction to communicate psychological pain and anguish. The cares of the Seafarer are 'hat ymb heortan' (11a), 'hot about the heart' and the Wanderer talks of binding his sorrowful mind with fetters (lines 19–21). In *The Seafarer*, thought disturbs the heart, which has an impact on the self and action (33b–7a):

> . . . cnyssað nu
> heortan geþohtas, þæt ic hean streamas,
> sealtyþa gelac sylf cunnige;
> monað modes lust mæla gehwylce
> ferð to feran.

> Now thoughts press upon the heart, that I myself should explore the high seas, the tumult of the salt-sea waves; the desire of the mind reminds the spirit to travel every time.

The mind has its own desire in this model of personhood, discrete from the desire of the self. Like this one, each of the elegies is structured on a self-reflective enunciation of personal feeling, informed by the application of remembered information to present circumstances and the mental outlook this produces. A dynamic between the past and the present is set up by which a change in outlook is indicated at the same time in which the expression of that outlook occurs. This dual time-frame is made possible by a separation of the mind from the self on the part of the speaker which allows the narrator to scrutinise his or her own state of mind, as shown above.

Culture and the construction of the self

The Old English elegies considered here are creative narratives which encode the values and ideas of the Anglo-Saxon poets who composed them, including culturally specific constructions of the self. Most significantly, the brief overview of the elegies offered above shows a high degree of interest in the self and in its dynamic psychological interaction with the mind in this literate society. Their narratives demonstrate the local belief in the possible

[138] Simon Nicholson, 'The Expression of Emotional Distress in Old English Prose and Verse', *Culture, Medicine and Psychiatry* 19 (1995), 327–38, at p. 337.

dislocation of self and mind, but also the imagined ideal of their coincidence. The poems encode the contemporary understanding of the self as the dominantly reflective individual identity, possibly engaged in a power struggle with the mind: a private not social self that is construed as undergoing turmoil and change, not one experiencing tranquillity and constancy.

Most of these attributes are strikingly familiar to us today. Once we as readers interpret the metaphoric and elegiac discourse, the self-contained though struggling selves depicted in Old English lament poems are immediately recognisable to us, despite the barrier of language and temporal distance. Perhaps this is why readers can relate to the sentiments of the verse today. Many features of the Anglo-Saxon self are typical, not of primitive or collective cultures, but of our contemporary ones: there is an awareness of the self with an interest in self-reform; a former self can be read as a text influencing the current version of the self; the self is a source of agency; and the self is a thinking, remembering entity involved in mental reorientation. Within this schema of the temporal development of the self, the literature enacts a psychological process whereby the present circumstances are reinterpreted by memory. The verse presents an energetic scrutiny of inner life as affective introspection, and imagines a self that requires renegotiation and management (contrary to the views of some modern psychologists, including Baumeister).

Just as modern psychological notions of the self-conceptualisations of earlier cultures can be revised in light of the evidence of Old English poetry, so too can our literary interpretations of those texts gain from some consideration of the ideas and methods from Cognitive Science. In Anglo-Saxon self-conceptualisation, the mind is one attribute of the individual, ideally though not always under the control of the self. These (frequently explicitly didactic) narratives are predicated on a belief in the shaping power of memory and contemplation, ultimately within the domain of the morally activated self. The poetry is presented as a process, not so much of self-discovery, but of mental realignment through a revisitation of the past and an acknowledgment of the role of the mind in the construction of the future. The poetry captures that particular construction of the self in its executive function of asserting its dominance over the mind and thereby its shaping power over individual reality. The self of the present is reoriented through this procedure of self-reflection that is grounded in the rehabilitating capacity of memory allied with logic. These poems depict an active and important role for the mind in the spiritual history of an individual and a key role for the self in personal moral consciousness. Although these ideas are clearly heavily indebted to Christian moralising rhetoric, they do not express any straightforward theological notion of subjective consciousness such as that articulated by Johnson and others. Rather, the secular themes of this poetry allow distinct views on identity and agency that are more locally specific to be introduced.

As we would expect in a culture as far removed from our present lives as that of medieval England, there are some crucial differences between Anglo-

Saxon constructions of the self perceptible in this poetry and those which are understood to be either universal or within a set of observed variables. The idea that the self can relinquish executive control to the mind or to some external force, and the possibility of the subjective dislocation required to allow this phenomenon to occur are not familiar models of the self today. Although the self is construed as a psychological entity in modern formulations, it is not perceived to be in competition with the mind as it is in Old English poetry. This manner of depicting an internal struggle comparable to the competing interests of dual allegiance, or conflicting hope and despair, or conflicted decision-making, allows an actual split in the self to be imagined, with implications for personal autonomy. The notion of divided identity is presented as a pressing dilemma requiring rational thought in order to mend the fragmentation of the self, perhaps with some analogy to the unification of multiple selves perceived as desirable in psychological therapy today.[139] The key difference is that the mind, not a separate self or selves, is construed as the complement entity in this medieval model of shifting power dominance within the complex of individual identity.

Aside from this deviation from our range of norms, the striking thing about the medieval construction of self evident from Old English lament poetry is how similar to recognised theorised models is the awareness of the self as an enduring source of identity, agency and cognitive functions in this culture. In particular, the degree of interest in the self is much higher than Cognitive and Social Scientists tend to acknowledge and the sophistication of thought far greater than is usually represented. Medievals did puzzle over the self and the nature of identity, and certainly used the evocative qualities of sophisticated vernacular poetry to express these ideas, and to explore the enduringly complex quandaries of consciousness, memory and thought-restraint. The mode of discourse and medium for discussing the self might have changed over time, but the desire to explore the self as a subject and the impact of cultural variables are as evident in this medieval context as in modern Western ones. This evidence strongly suggests the value of greater diachronic depth to the study of the self and identity among cognitive scientists, and a more rigorous challenge to the notion amongst them and many scholars in the humanities that introspection is particularly tied to early-modern and modern cultures. Rather, as Old English poetry shows, introspection, mind-reading and indeed the representation of emotion, are integral features of pre-modern literature, as is its capacity to invite an affective response in its audience, the subject of the next chapter.

[139] See, for instance, the work of Danziger, 'The Historical Formation of Selves'; McAdams, 'The Case for Unity'; and Rosenberg, 'Multiplicity of Selves'.

7

Cognitive Approaches to the History of Emotions and the Emotional Dynamic of Literature

The history of emotions and cross-cultural intelligibility

Emotion, and its connection to cognitive functioning, has come up again and again in the preceding chapters, which is unsurprising given the reliance of poetry on affective experience. Wordsworth famously observed that poetry has emotion as its essence: 'All good poetry is the spontaneous overflow of powerful feelings: it takes its origin from emotion recollected in tranquillity'.[1] More recently, cognitive scientists have begun to analyse with their own concepts and methods this uncontested capacity of literary texts to represent, simulate and cause emotions,[2] thereby facilitating a dynamic line of inquiry for the cognitive study of literature. Any consideration of this combination of emotion and poetry, or literature more broadly, has two related dimensions, both of which are essentially cognitive and can be analysed via specifically cognitive approaches: how emotion is represented in literature, and how literary texts can trigger emotional reactions in readers. When we consider texts created or at least written down over a thousand years ago, as in the case of Old English poetry, further considerations arise from these two intersections: is the experience being represented in these medieval texts the same as similarly named emotions experienced by us today; and more broadly, to what degree are emotions intelligible cross-culturally. To explore these issues, the discussion below, pp. 170–5, revisits two texts already treated above – *Wulf and Eadwacer* and *Beowulf* – from a new perspective, the History of Emotions.

The History of Emotions is a field experiencing a great surge of interest at present, indicated by the emergence of publications[3] and research centres.[4]

1 Preface 1800 version (with 1802 variants) to the *Lyrical Ballads*, 2nd edn, ed. R. L. Brett and A. R. Jones (London, 1991), pp. 233–58, at p. 251.

2 The work of Keith Oatley is foremost in this field: see, for example, 'Simulation of Substance and Shadow: Inner Emotions and Outer Behavior in Shakespeare's Psychology of Character', *College English* 33.1 (2006), 15–33, where he argues that in audiences of Shakespeare's plays, emotion is instantiated: 'the words and emotions become the mind of the reader' (30); and Oatley, *Emotions: A Brief History* (Malden, MA, 2004), esp. pp. 6 and 13, where the interrelationship of emotion and literature is asserted as a fundamental truth.

3 For example, in Cultural History: Robert A. Kaster, *Emotion, Restraint and Community in Ancient Rome* (Oxford and New York, 2005); and Barbara H. Rosenwein, *Emotional Communities in the Middle Ages* (Ithaca and London, 2006). Note also the new series from Oxford University Press, Emotions of the Past, gen. eds Robert Kaster and David Konstan.

4 Notably, the Australian Research Council Centre of Excellence in the History of Emotions – Europe 1100–1800, based at the University of Western Australia under the directorship

While this field brings scholars from several fields together, most of them are from disciplinary backgrounds in the humanities, and treat emotion from an historical perspective that draws very little, if anything, from Cognitive Science. Rather, scholars mine historical and literary texts for representations of the emotional life and for evidence of how emotions change over time. Similarly, the interest paid to the subject of emotions by Cognitive Science is restricted, where scholars are rarely interested in the diachronic aspect of the affective life, and have only relatively recently become interested in the subject at all in a sustained and energetic manner, including affective responses to the arts.[5] Given the distinctly separate nature of these lines of research, the nature of the topic, and the evidence available in the documents of earlier cultures provides a particularly fruitful potential for more actively interdisciplinary inquiry in this area. The textual capacity both to represent the flux of emotional states and also to trigger an emotional reaction in the reader are matters of keen interest both to cognitive scientists and to cultural historians that can be brought into closer dialogue through a specifically cognitive approach.

Emotions are the subject of a great deal of interest in the many disciplines captured by the phrase Cognitive Science – in Philosophy, Psychology, Anthropology, Sociology, Cognitive Neuroscience – and beyond.[6] In Cognitive Science, an emotion is commonly defined as 'a psychological state or process that functions in the management of goals', a state 'seen to serve important intracognitive and interpersonal functions'.[7] The concept of 'management' in these statements captures the foundational role of emotions in determining the entire quality of lived experience. Many cognitive psychologists, like Keith Oatley, favour this 'appraisal' theory to account for the source of emotions. This essentially functionalist theory holds that emotions are caused not by events, but by 'appraisal of events in relation to goals and plans, an idea that has found its way into cognitive studies of literature.[8] Similarly, in Neuroscience emotions are seen as 'brain states and bodily responses'[9] that

of Philippa Maddern, funded to an extent unheard of in the humanities, of over $24.25 million over seven years. Further details at: http://www.emotions.uwa.edu.au/ . See also the partner institution, the centre at Queen Mary, University of London: http://www.qmul.ac.uk/emotions/.

5 Oatley, *Emotions: A Brief History*; and Jenefer Robinson, *Deeper than Reason: Emotion and its Role in Literature, Music, and Art* (Oxford, 2005). For the comparative linguistic point of view, see Anna Wierzbicka, *Emotions across Language and Cultures: Diversity and Universals* (Cambridge, 1999).

6 The situation has changed since Joseph LeDoux's observation that most cognitive scientists avoid the study of emotions. See *The Emotional Brain: The Mysterious Underpinnings of Emotional Life* (New York, 1998), pp. 33–5.

7 Keith Oatley, 'Emotions', in *MIT Encyclopedia of the Cognitive Sciences*, ed. Robert A. Wilson and Frank C. Keil (Cambridge, MA, 1999), pp. 273–5, at p. 273.

8 Lalita Pandit, 'Emotion, Perception and Anagnorisis in *The Comedy of Errors*: A Cognitive Perspective', *College Literature* 33.1 (2006), 94–126, at p. 95. For a challenge to the appraisal view, see Patrick Colm Hogan, 'On Being Moved: Cognition and Emotion in Literature and Film', in *Introduction to Cognitive Cultural Studies*, ed. Lisa Zunshine (Baltimore, 2010), pp. 237–56.

9 LeDoux, *The Emotional Brain*, p. 302.

act as 'managers of mental life, prompting heuristics that relate the flow of daily events to goals and social concerns'.[10] These conceptions of emotions as managers or regulators whose remit extends beyond the life of mere feelings are especially interesting to the literary or cultural historian, as they underscore the interpenetration of the emotional life with cognitive functions, consciousness and an enduring sense of the self. Research into the emotions from several fields has shown us that though we might recognise apparently separate emotions, such as happiness or anger, the emotional web is tangled and interconnected with other aspects of being, sensing and interacting.

Given the size of the topic, scholars quite reasonably restrict their scope of inquiry in order to study a manageable slice of the emotional life. For example, major streams within Psychology include the study of the way in which emotions are at the foundation of human motivational systems;[11] the distinction between emotions and feelings;[12] and between an emotional reaction and an empathetic response. Pragmatically, scholars in Psychology recognise seven basic emotions: fear, anger, disgust, sadness, joy, shame and guilt,[13] though the list and the number of items on it changes depending on who is consulted, and many scholars distinguish primary from secondary or 'social' emotions.[14] Furthermore, cross-cultural studies have shown how difficult the terminology around the emotions is in relation to semantic range and translation across cultures.[15]

Many of the emerging orthodoxies from these disciplinary studies of emotion are useful to scholars of literature and history. For instance, sociologists have long argued that social experience conditions the processing of

10 Oatley, 'Emotions', p. 275.

11 Silvan Tompkins is the scholar best known for this line of thought, part of a theory of personality based on the idea of scripts underpinned by emotions: see *Exploring Affect: The Selected Writings of Silvan S. Tomkins*, ed. E. Virginia Demos (Cambridge, 1995).

12 Jesse Prinz, 'Are Emotions Feeling?', *Journal of Consciousness Studies* 12.8–10 (2005), 9–25.

13 Morton L. Kringelbach, 'Emotions, Feelings, and Hedonics in the Human Brain', in *The Emotions: A Cultural Reader*, ed. Helena Wulff (Oxford and New York, 2007), pp. 37–60, at p. 41.

14 Antonio Damasio includes on this list: embarrassment, jealousy, guilt and shame, and lists six primary emotions, with surprise in the place of guilt and shame in the list above. See *The Feeling of What Happens: Body and Emotion in the Making of Consciousness* (Orlando, 1999), p. 51. He also posits the idea of 'background emotions' such as 'well-being or malaise, calm or tension' (p. 52).

15 The literature is huge, and dominated by studies of particular groups of words and specific cultural contrasts. For the Cognitive Semantics view, see the articles collected in *Happiness: Cognition, Experience, Language*, ed. Heli Tissari, Anne Birgitta Pessi and Mikko Salmela, Collegium Studies across Disciplines in the Humanities and Social Sciences (Helsinki, 2008), at: http://www.helsinki.fi/collegium/e-series/volumes/volume_3/index.htm; and Zoltán Kövecses, 'Metaphor and Thought', in *The Cambridge Handbook of Metaphor and Thought*, ed. Raymond W. Gibbs (Cambridge, 2008), pp. 380–96. For an Old English example, see Dirk Geeraerts and Caroline Gevaert, 'Hearts and (Angry) Minds in Old English', in *Culture, Body, and Language: Conceptualizations of Internal Body Organs across Cultures and Languages*, ed. Farzad Sharifian, René Dirven, Ning Yu, and Susanne Niemeier (New York and Berlin, 2008), pp. 319–47. For a broader overview of key concepts and readings in Cultural Studies, see: *Emotions: A Cultural Studies Reader*, ed. Jennifer Harding and E. Deidre Pribram (London, 2009).

emotions,[16] and have settled on the culturally specific experience of emotions, and therefore 'how mistaken it would be to assume that emotions are unproblematically translatable from one culture or historical period to another'.[17] Similarly, cultural anthropologists insist upon the constitutive role of social context in the very experience of emotions, and have increasingly emphasised the discursive resources of a community as a factor: 'the most productive analytical approach to the cross-cultural study of emotion is to examine discourses on emotion and emotional discourses as social practices within diverse ethnographic contexts'.[18]

Despite this interest in culture and emotion, scholars in the Social Sciences rarely consider literary texts as sources of information on the emotional life, even when considering emotions as 'psychosocial' phenomena in social life and popular culture,[19] though there are some notable exceptions. Keith Oatley and his colleagues, for instance, briefly discuss the ubiquitous textual focus on the emotions: 'written narrative literature, from ancient times to the present, concentrates on our emotional lives... Publicly available stories give members of society common exemplars of action of emotion and of responsibility'.[20] Similarly, the moral philosopher Martha Nussbaum approaches literature as a site in which philosophical and religious views on emotions are represented, in her model in which there is an ethical dimension to appraisal.[21] On the other hand, there is a lack of cross-disciplinary inquiry from the other direction too: scholars working in the History of Emotions often ignore Cognitive Science altogether, focusing instead on philosophical or theological ideas.[22]

Scholars working in the hard sciences are even less likely to be interested in textual representations of and triggers for emotions, though literary scholars can certainly deploy some emerging ideas from these areas. For instance, recent work in Neurobiology (most notably by Antonio Damasio) has shown that the emotional life is thoroughly embedded in and interacts with cognitive functioning and embodied experience, confirming what William James postulated in 1884 (that emotions are the perceptions of changes in the body).[23]

16 Cas Wouteres, 'The Civilizing of Emotions: Formalization and Informalization', in *Theorizing Emotions: Sociological Explorations and Applications*, ed. D. Hopkins, H. Kuzmics, H. Flam and J. Kleres (Frankfurt and New York, 2009), pp. 169–93.

17 Robert A. LeVine, 'Afterword', *The Emotions: A Cultural Reader*, ed. Wulff, pp. 397–9, at p. 398.

18 Lila Abu-Lughod and Catherine A. Lutz, 'Introduction: Emotion, Discourse, and the Politics of Everyday Life', in *Language and the Politics of Emotion*, ed. Lutz and Abu-Lughod (Cambridge, 1990), pp. 1–23, at p. 1.

19 See, for example, S. D. Sclater, D. W. Jones, H. Price and C. Yates, eds, *Emotion: New Psychosocial Perspectives* (London, 2009), 'Part II: Emotions in the Public Sphere'.

20 Keith Oatley, Dacher Keltner and Jennifer M Jenkins, *Understanding Emotions*, 2nd edn (Oxford, 2006), p. 401. See also Keith Oatley, 'A Taxonomy of the Emotions of Literary Response and a Theory of Identification in Fictional Narrative', *Poetics* 23 (1994), 53–74.

21 Martha C. Nussbaum, *Upheavals of Thought: The Intelligence of Emotions* (Cambridge, 2001), esp. pp. 457–613. Nussbaum's work, though influential, has been the subject of serious criticism. See, for example, Dorothy J. Hale, 'Aesthetics and the New Ethics: Theorizing the Novel in the Twenty-First Century', *Publications of the Modern Language Association of America* 124.3 (2009), 896–905.

22 For example, Ramsay MacMullen, *Feelings in History, Ancient and Modern* (Claremont, 2003); and Simo Knuuttila, *Emotions in Ancient and Medieval Philosophy* (Oxford, 2004).

23 Damasio, *The Feeling of What Happens*, pp. 287–8.

Furthermore, once it is granted that emotions are determined biologically, the implication follows that the evolved human brain behaves consistently and even automatically when it comes to emotions, notwithstanding some cultural variations: 'the considerable amount of individual variation and the fact that culture plays a role in shaping some inducers do not deny the fundamental stereotypicity, automaticity, and regulatory purpose of the emotions'.[24] Given this orthodoxy, many neuroscientists are currently investigating the brain functioning of emotion, and increasingly place emphasis on the role of memory in the process.[25] In particular, the reliance of consciousness on working memory implicates the role of emotions: 'feelings result when working memory is occupied with the fact that one's brain and body are in a state of emotional arousal'.[26] In turn, the study of human memory and consciousness is attracting a growing range of scholarly inquiries within Cognitive Literary and Cultural Studies too, which might also come to integrate research on human emotion more fully.

Approaches to emotion and literature from outside Cognitive Science that are nevertheless avowedly 'cognitive' are seeking to broach this divide between cognition and culture. They consider how the functioning of the relatively stable evolved biological embodied brain and its emotional life interacts with its culturally contingent set of circumstances. In other words, these approaches consider both how cognitive functioning shapes culture, and also how cognition is culturally determined, an influence that even the neuroscientists acknowledge ('learning and culture alter the expression of emotions and give emotions new meanings'[27]). Further, these humanities scholars consider how cognition and emotion are deeply interrelated, and how literary texts can model that interaction, and indeed embody cognitive structures.[28] More broadly, there is now some consideration of the emotional pull of the literary text, of how 'fiction gives people the chance to practice their emotional connections with other people'.[29] So, scholars working at the interdisciplinary nexus of science, Social Science and the arts are coming

24 Ibid., p. 51.

25 'Emotion has important effects on mental functions that are indisputably cognitive, such as memory, attention, and perception'. See Richard D. Lane, Lynn Nadel, John J. B. Allen and Alfred W. Kasniak, 'The Study of Emotion from the Perspective of Cognitive Neuroscience', in *Cognitive Neuroscience of Emotion*, ed. Richard D. Lane and Lynn Nadel (Oxford, 2000), pp. 3–11, at p. 4.

26 Joseph LeDoux and Michael Rogan, 'Emotion and the Animal Brain', in *MIT Encyclopedia*, ed. Wilson and Keil, pp. 269–71, at p. 270.

27 Damasio, *The Feeling of What Happens*, p. 51.

28 Lalita Pandit and Patrick Colm Hogan, 'Introduction: Morsels and Modules: On Embodying Cognition in Shakespeare's Plays', *College English* 33.1 (2006), 1–13; Evelyn Tribble, *Cognition in the Globe: Memory and Attention in Shakespeare's Theatre* (London, 2011); and Evelyn Tribble and Nicholas Keene, *Cognitive Ecologies and the History of Remembering: Religion, Education, and Memory in Early Modern England* (London, 2011).

29 Blakey Vermeule, *Why Do We Care about Literary Characters?* (Baltimore, 2010), p. 165. Vermeule's argument bears some similarity to that of Lisa Zunshine, *Why We Read Fiction: Theory of Mind and the Novel* (Colombus, 2006), p. 164: 'Fiction helps us to pattern in newly nuanced ways our emotions and perceptions', though the latter's main focus is Theory of Mind rather than the emotions.

to appreciate the rich potential of considering the complex inter-relance of culture, cognition and emotion.

Within this burgeoning cross-section, there is considerable interest in whether we engage with representations of emotions in the same way as we do with real-life emotional encounters. In a specifically cognitive approach, Patrick Colm Hogan, argues against the appraisal theory, in his discussion of emotions experienced in fictional encounters.[30] Emotional reactions to fiction, Hogan argues, use the same sort of neurologically created 'concrete images' and 'emotional memories'[31] that determine real-life emotional responses. He recruits ideas from Neurobiology to argue that imagination, when engaged in reading or viewing a text, creates a distance between the 'emotional object' represented in that text and 'the egocentric space of direct action'[32] – the range of proximity that triggers a direct response. In other words, the reading brain responds to perceived spatial distance, senses a buffer zone, that allows us to experience an emotion, but less directly than in real life, and with some degree of cultural variation. Gregory Currie, on the other hand, presents a different solution to the problem of how emotions are engaged via fiction, but who likewise accounts for the apparent distance between emotional trigger and reaction. He suggests that a virtual reader is created in the mind of the reader, a 'reader of fact', who reacts to the characters and events of the narrative, whose perspective we use as a guide for emotional interaction.[33]

Scholars who study emotion and fictionality often focus on popular culture, especially film. Noël Carroll, for instance, explores the 'affective address of popular fiction',[34] and Amy Coplan examines the distinctive capacity of film to trigger genuine emotional reactions.[35] Both scholars emphasise the powerful emotional triggers of audiovisual cues, especially facial expressions, and come up with different explanations for the viewers emotional response: Carroll denies that consumers of fiction simulate the emotional experience of characters as some scholars argue, but rather 'mobilize an affective stance' that is distinctly produced in response to a fiction whose content we imagine rather than believe.[36] Coplan prefers the theory of 'emotional contagion', whereby an automatic affective response is produced when we observe the experience of emotion in others, a phenomenon requiring the sensory input of a visual experience as occurs when viewing a film; literary fictions, on the other hand, can produce reactions that are more cognitive, but less affective because they produce experiences more removed from real-world sensations.[37]

30 Hogan, 'On Being Moved'.
31 Ibid., p. 246.
32 Ibid., p. 248.
33 Gregory Currie, 'The Paradox of Caring: Fiction and the Philosophy of Mind', in *Emotion and the Arts*, ed. Mette Hjort and Sue Laver (Oxford, 1997), pp. 63–77, at p. 71.
34 Noël Carroll, 'On the Ties that Bind: Characters, the Emotions, and Popular Fictions', in *Philosophy and the Interpretation of Pop Culture*, ed. William Irwin and Jorge J. E. Gracia (Lanham, 2007), pp. 89–116, at p. 89.
35 Amy Coplan, 'Catching Characters' Emotions: Emotional Contagion Responses to Narrative Fiction Film', *Film Studies* 8 (2006), 26–38.
36 Ibid., p. 91.
37 Ibid., p. 35

Coplan, like other scholars working in this area, distinguishes emotional contagion from empathy, 'a complex and unique imaginative process involving both cognition and affect',[38] which is why literary narratives can produce empathy.[39] Empathy, or 'a vicarious, spontaneous sharing of affect', can be created through character identification and emotional simulation, primarily via the activation of mirror neurons,[40] which function pragmatically to assist in the understanding of the actions and emotions of others. Mirror neurons allow us to experience the emotional meaning of our interpersonal encounters – and perhaps also experience empathy – by producing 'reflexive processing of the sensory aspects linked to how they appear in the facial expressions or acts of others'.[41] Clearly, imagination can produce images in the brain of facial expressions that in turn trigger the same sort of emotive mirroring response. Fritz Breithaupt, in his fuller treatment of this area, covers four main models of empathy, including one from narratology, concluding that 'narrative fiction can only exist because it invites, triggers, channels, controls, and manages empathy'.[42] Other theories of affective response to literary fictions often rely on one of Breithaupt's models, Simulation Theory. For instance, Meskin and Weinberg cite the orthodoxy that 'engaging with fiction is a matter of "off-line simulation"', before offering their own modification that relies on the idea that fictions trigger the creation of possible worlds in the mind of the recipient.[43]

This idea bears similarity to the initially linguistic approach to the imaginative creation of text worlds, developed in Cognitive Linguistics and treated in Chapter 4 above, that people make sense of discourse through the creation of mental representations of the ideas provoked by that discourse. Going beyond that idea to a more nuanced consideration of the role of emotion in the aesthetics of reading is Peter Stockwell's theory of 'texture': 'the experienced quality of textuality'.[44] Stockwell develops a highly technical analytical means of analysing linguistic cues that invite the build-up of emotional response to a text, but more immediately useful here is his central contention that readers of literary texts can have a cognitive affective response to that text that is susceptible to systematic scrutiny. This more literary development of Text World Theory is particularly useful in that it seeks to account for the mechanisms of literary reading: how certain ideas and knowledge from a reader's whole store of information are selectively recalled and deployed in response to textual cues in a particular act of reading; and how texts

38 Ibid., p. 31

39 For a good overview on philosophical perspectives on readers' empathetic engagement with fictional narratives and characters, see Amy Coplan, 'Empathetic Engagement with Narrative Fictions', *Journal of Aesthetics and Art Criticism* 62.2 (2004), 141–52.

40 Suzanne Keen, 'A Theory of Narrative Empathy', *Narrative* 14.3 (2006), 207–36, at p. 208.

41 Giacomo Rizzolatti and Corrado Sinigaglia, *Mirrors in the Brain – How Our Minds Share Actions and Emotions*, trans. Frances Anderson (Oxford, 2008), p. 190.

42 Fritz Breithaupt, 'How is it Possible to Have Empathy? Four Models', in *Theory of Mind and Literature*, ed. P. Leverage, Howard Mancing, Richard Schweickert and Jennifer Marston William (West Lafayette, 2011), pp. 273–88, at p. 274.

43 Aaron Meskin and Johathan M. Weinberg, 'Emotions, Fiction, and Cognitive Architecture', *British Journal of Aesthetics* 43.1 (2003), 18–34.

44 Peter Stockwell, *Texture: A Cognitive Aesthetics of Reading* (Edinburgh, 2009), p. 1.

foreground and group certain ideas, inviting certain associations and cumulative responses in the mind of the reader in the creation of an emotional response to it. Specifically, Stockwell has developed a stylistic approach that accounts for the literary creation of ambiguity, tone and the affective reading experience. He looks at how the reading mind is primed below the level of consciousness to create certain connections in the text at hand and to derive a particular sort of reading experience from that text.[45] Even though Stockwell's method is intricately based on close linguistic analysis, its conceptual principles have broader value beyond the field of Cognitive Linguistics, particularly his idea of 'the texture of edges', or 'the flow of experience across a transitional moment'.[46] By this remark he means that the human mind is primed to notice changes and to perceive edges, and movement across boundaries, both physically and metaphorically. When texts represent changes in emotional states, the joining edge is particularly noticeable, and can be a powerful emotional trigger itself.

With these ideas in mind, it would be useful to consider how a modern reader can have an emotional reaction to a medieval text, in combination with an intellectual and aesthetic experience produced by that literary encounter. An approach to this complex issue could be developed by blending ideas and methods from Cognitive Science, Cognitive Poetics, Cultural History and Literary Analysis. In particular, this approach would need to combine insights from neuroscientific work on the role of emotion in human mental processes with literary approaches to the role of emotion in the aesthetics of reading. Old English poetry lends itself to such an approach, as the Anglo-Saxon tendency to refer to the mind rather than to the person as the site of emotion and the recipient of fate suggests not just a distinctly different conscious way of viewing human perception and the outcome of events from our own, but also an unconscious, culturally determined mind schema that is both alien to us and yet sufficiently recognisable for the text to make sense to us, albeit via a process of translation. The Anglo-Saxon literary corpus, and in particular its fictional representations of consciousness, are readable to us, but only via the process of linguistic and cultural relocation that operates through translation into modern English. Nevertheless, a core degree of intelligibility remains, stemming from our evolutionary proximity, and because we share the human experience of an embodied mind and, apparently, a hard-wired predisposition for narrative. The similarity of our emotional experiences with other humans from remote culture makes cross-cultural intelligibility possible, as discussed above. Our use of conceptual metaphor to communicate abstract ideas is a key point of contact with our medieval predecessors, whose literary representations of embodied emotions we are able to recognise and appreciate.

The literature of Anglo-Saxon England has been analysed with respect to emotions, but not from a particularly cognitive perspective. Simon Nicholson,

45 Ibid., p. 11
46 Ibid., p. 107

for instance, has studied the abstract nouns connoting moods in Old English literature, and concluded that there is evidence in the corpus of both 'psychological and somatic expressions of emotional distress', thereby disputing the idea that such a distinction is relatively recent.[47] Other studies have been similarly mainly lexical.[48] But the archaeology of emotions in Anglo-Saxon England, and the capacity of its literature to produce an emotional engagement, even an empathetic response, is as yet little explored. More broadly, research into the degree to which emotions are culturally induced or shaped, and the specific categorisation of emotional states, can be advanced by a consideration of the emotional content and expression of Old English verse. Because the quality of emotional experiences is central to conceptual history and psychological considerations of questions such as 'do emotions change over time?', the fullest diachronic range available should be deployed in any attempt to address these large issues. The Anglo-Saxon contribution to Social and Cognitive History is yet to be written, but to indicate how we might start to conceive of such a project, I will briefly suggest below how these ideas, particularly the shared cognitive basis of meaning and feeling, might be approached in relation to medieval literary texts. I take as my examples extracts from two very different Old English poems: the enigmatic *Wulf and Eadwacer*, and the heroic narrative *Beowulf*.

The emotional dynamic of Wulf and Eadwacer

The short poem *Wulf and Eadwacer*, already cited in full and discussed above (pp. 95–102) in relation to Text World Theory, is an allusive (and interpretively elusive) conglomerate of emotional expressions. Modern readers have found this text enigmatic as a narrative (in terms of the relationship among the speaker, and the two named characters), but the expression of emotions

47 Simon Nicholson, 'The Expression of Emotional Distress in Old English Prose and Verse', *Culture, Medicine and Psychiatry* 19 (1995), 327–38, at p. 337.

48 Edwin N. Gorsuch, 'Emotional Expression in a Manuscript of Bede's *Historia Ecclesiastica*: British Library Cotton Tiberius A XIV', *Semiotica* 83 (1991), 227–49; Edgar C. Polomé, 'Some Comments on the Vocabulary of Emotion in Germanic', in *Interdigitations: Essays for Irmengard Rauch*, ed. Gerald F. Carr, Wayne Harbert and Lihua Zhang (Bern, Frankfurt am Main and New York, 1998), pp. 129–40; Malgorzata Fabiszak, 'A Semantic Analysis of Emotion Terms in Old English', *Studia Anglica Posnaniensia* 34 (1999), 133–46; Michiko Ogura, 'Verbs of Emotion with Reflexive Constructions', in *Lexis and Texts in Early English: Studies Presented to Jane Roberts*, ed. Christian Kay and Louise Sylvester (Amsterdam and Atlanta, 2001), pp. 203–12; Hans-Jürgen Diller, 'The Growth of the English Emotion Lexicon: A First Look at the *Historical Thesaurus of English*', in *Of Dyuersitie & Chaunge of Langage: Essays Presented to Manfred Görlach on the Occasion of His 65th Birthday*, ed. Katja Lenz and Ruth Möhlig, Anglistische Forschungen 308 (Heidelberg, 2002), pp. 103–14; Michiko Ogura, 'Words of EMOTION in Old and Middle English', in *A Changing World of Words: Studies in English Historical Lexicography, Lexicology and Semantics*, ed. Javier E. Díaz Vera, Costerus n.s. 141 (Amsterdam and Atlanta, 2002), pp. 484–99; Michiko Ogura, 'Words of Emotion in Old and Middle English Translations of Boethius's *De Consolatione Philosophiae*', in *Text and Language in Medieval English Prose: A Festschrift for Tadao Kubouchi*, ed. Akio Oizumi, Jacek Fisiak and John Scahill, Studies in English Medieval Language and Literature 12 (Frankfurt am Main, 2005), pp. 183–206; Michiko Ogura, 'Old and Middle English Verbs of Emotion', *Poetica* 66 (2006), 53–72; and Hans Sauer, 'Ælfric and Emotion', *Poetica* 66 (2006), 37–52.

is not difficult to grasp, especially in the context of other Old English lyric texts, as discussed above. In several Old English elegiac narratives, a speaker laments his or her separation from the group,[49] though here, separation from an individual, presumably a lover, is articulated in lines 3–4. The conceptual metaphor of the self as an island is used here to express the sadness of isolation and loss. The emotional trajectory moves then to the registration of fear and anxiety, then to joy and disgust at the memory of (presumably) an act of love that was later regretted in shame, then on to sadness, fear and anxiety again, amplified by the potential connotations of the guilt attendant on the loss of a child, 'uncerne earne hwelp' (16b), 'our wretched cub'. The second half of the poem in particular is a self-aware expression of a complex emotional state, the conflicted nature of which is registered physically. Embodied emotion, or emotions as perceptions of changes in the body, is expressed in the line about sickness, and conflicted bitter-sweet emotions in the line about joy and loathing (13). Regret mixed with resignation is expressed in the final paradoxical statement about separation in togetherness. Most interestingly, though, is the degree of self-awareness of the quality and impact of the emotions being experienced, with implications for the assumed communicative and social function of the text.

Even though it might refer to social or historical information that is lacking to the modern audience, the text is unambiguously a lament in which human emotional pain of a recognisable nature is encapsulated in language. The complex, experiential nature of the emotions represented seems familiar, and highly cross-culturally intelligible. The recognised emotions allow us to create and to fill out the text world here, and to notice particularly the emotional shifts from anxiety, joy to sorrow, to longing, to resignation, and also the manner in which emotional states are embedded within other states (for instance, joy within sorrow, resignation within regret and the reverse). Neurologically, the overriding emotion of grief, and the conflicted sense of joy and grief in memory, dominates consciousness in this representation, thereby managing the mental life of the speaker and controlling the flow of daily events towards the goal of indulged memory of a loved one – the communicative rationale of the text. Its social functions, besides the aesthetic expression of the lament, might include the reinforcement of the social status quo, through depiction of the consequences of transgression. In this manner,

[49] The narrator of *The Wife's Lament* grieves for her lover and their usual social context, and the speaker of *The Seafarer* mourns his alienation at sea. In *The Wanderer*, the ritual gift-giving ceremony in the hall is recalled, along with the speaker's affection for his lord. The remembered self is 'sociocentric' in that social norms and responsibilities guide behaviour, though this is contrasted with the present circumstances of the solitary and self-oriented narrator. See E. Tory Higgins, 'The "Self-Digest": Self-Knowledge Serving Self-Regulatory Functions', *Journal of Personality and Social Psychology* 71.6 (1996), 1062–83. The view of the self shaped by emotional experience here is independent according to the formulation of Markus and Kitayama: the emotions, especially grief, are ego-focused, and other selves are perceived to be outside the bounds of the self rather than connected to it. The poems frequently depict this personal isolation by means of the metaphor of physical isolation. Hazel Rose Markus and Shinobu Kitayama, 'Culture and the Self: Implications for Cognition, Emotion, and Motivation', *Psychological Review* 989.2 (1991), 224–53.

the text has the potential to trigger the emotion of fear, via the arousal of compassion for the distraught speaker and her expression of loss and sadness.

The shape and force of the emotional trajectory of this text, and its emotional impact, seem very familiar, and draw the modern-day reader towards, not away from, the Anglo-Saxon one. Partly, that conformity stems from the causes of the emotions depicted here. It is worth noting that all seven of the basic emotions are recognisable in this poem – fear, anger, disgust, sadness, joy, shame and guilt – with sadness playing the keynote. It is not therefore surprising to come away from the poem feeling that the full gamut of experience has been run, and to have noticed the shifts and boundary-crossings that the interpreter has cognitively incurred in making sense of it. If we consider an emotion as an ephemeral psychological state that both drives motivation and manages immediate goals, then it is useful to see how the representation of the affective life here shows the particular organisation of the inner life around emotions in general and sadness in particular. The articulation of this experience in poetic form can in turn be viewed as the entrenchment of this phenomenon rather than an alleviation of it, and indeed, a vehicle by which this emotional perspective can be spread.

The poem enacts the now common idea that 'emotions can distort practical reasoning', particularly during 'experiential imagining', imagining from the inside.[50] Of course, the reader or hearer of the text is invited to sympathise with the speaker via the simulation of the emotions being expressed. This activity occurs through the interpretive act of experiential imagining necessary in the act of textual sense-making, so to a limited degree the audience experiences the emotions too. But, given the distance created via the fictionalising process, the audience remains in control of practical reasoning, and is metacognitively aware of its distorting effect on the speaker, something that is indeed highlighted by the allusive, enigmatic narrative arc of this lament. Because the modern reader is left wondering who has done what to whom, the emotional texture rather dominates, not the precise details, in this act of 'off-line simulation'. Within that ambiguity, though, are sufficiently recognisable human emotional high-water marks to signal personal agony: the emotional acuteness of the mother–child separation and the longing for an absent lover depicted in this poem particularly captures the dramatic apex of its emotional landscape, and the impersonal constructions encode the universal quality of the tragedy recounted.

Given the emphasis on grief and longing in this text, one might ask: What does this text tell us about the Anglo-Saxon apprehension of psychological pain, and how is that apprehension similar to or distinct from our own? Our understanding and empathy with the emotional complexity of the subjective experience represented in this text are predicated on comparable human psychological functioning shared by the creator and receiver of this text. Moreover, the apparent cross-cultural intelligibility of the text and its

[50] Peter Goldie, 'Imagination and the Distorting Power of Emotion', *Journal of Consciousness Studies* 12.8–10 (2005), 127–39, at pp. 127–8.

emotional texture points to consistency rather than variation in human apprehension of and cause for emotional pain, as well as the deep entrenchment of the reliance on poetry to represent and to engage with the emotional life. Indeed, the intensely affective quality of this poem is integral to its narrative ambiguity and capacity to evoke an emotional response in its audience. This affective power derives in part from the investment required on the part of the audience, and on the capacity of poetry for emotional co-experience. More specifically, it draws on the acknowledged role of emotions as managers of the inner life and controllers of cognitive functioning, and on the capacity of poetry to behave as emotion realised in aesthetic form.

The audience has to embrace these concepts in the act of sense-making, and thereby to valorise the emotional quality of lived experience, particularly highly charged in this fictional representation. The patterns of emotional behaviour depicted in the text are mimicked by the reader in the act of interpretation, and so experienced in some avatistic manner in this mobilisation of an affective stance towards the text that is cognitive as well as emotional. Of such experience, empathy is born. That readers today can feel such empathy, and live through the emotions depicted in *Wulf and Eadwacer*, tells against a model of the emotions that relies heavily on their cultural construction. No doubt historical and social forces have some impact on emotional experience, but the cross-cultural intelligibility of the emotional power of poetry argues for a limit on that shaping influence. Moreover, the emotional logic and contoured affective landscape of this poem shows the durable aesthetic reliance on the individual inner life, and the dominant continuities in the specifics of embodied human emotional experience over the millennia.

The emotional life in Beowulf

Similarly, the empathetic engagement of the audience is invited in *Beowulf*, an intuitively as well as demonstrably emotional text.[51] There are many instances of emotional concentration in this long text: speeches, notably by Hrothgar and Beowulf, but on many other occasions too; in representations of distributed emotion – when entire communities are said to respond to events with a communal emotion; in individual reactions to catastrophe and tragedy; and the recounted expression of grief in laments. There is enormous scope to approach this text via the emotions, but the following discussion will indicate how that line might be pursued via a few brief examples. For instance, notwithstanding the negative depiction of the monsters in this text in their role as adversaries to the focal hero, there are moments where the audience is invited to sympathise with them and thereby to develop an affective stance that complicates the narrative alignment of the reader's support for and focus on the hero.

[51] See the statistical psychologically based charting of emotional activity represented in the poem: Cynthia Whissell, 'The Flow of Emotion through Beowulf', *Psychological Reports* 99 (2006), 835–50.

For example, Grendel's mother is initially and fundamentally motivated by grief. Her maternal anguish at the sight of her dead son's arm provides a brief humanising quality to her and a pathetic aspect to her behaviour. Indeed, when the audience is first introduced to her, it is to see her grief-driven desire for vengeance: 'Grendles modor, / ides aglæcwif yrmþe gemunde' (1258b–1259), 'Grendel's mother, a woman, female monster, mourned her misery'. Then, 'heo under heolfre genam / cuþe folme; cearu wæs geniwod, geworden in wicun' (1302b–1304a), 'she took the known hand covered with blood; care was renewed, had come to pass in the dwellings'.[52] The limb was referred to earlier as 'blodge beadufolme' (990a), 'bloody battlehand' (which caused the blood of others to flow), but now her emotional reaction is embedded in the 'known hand' (known through 'blood' kinship': the conceptual metaphor of 'blood' is working hard here). The *Apo Koinou* line can refer backwards as well as forwards: the renewal of care can refer both to her renewed grief and to the fresh misery she inflicts on those nearby when she attacks. Like Grendel, his mother is humanised to a sufficient degree for their behaviour to be seen as calculated and evil rather that merely instinctive and bestial. Grief here is the trigger for motivational systems – it is managing behaviour, activating memory, and prompting revenge – and seems to determine rather than to disable practical reasoning, and to singularise the agent's goals.

The reader or hearer of this text must engage with this emotional territory in order to make sense of the narrative, and thereby invests the monsters with human emotional characteristics in order to conceptualise motivated behaviour.[53] This conceptual activity is reinforced by the text's invitation to see the loss as psychologically distressing on both sides: 'ne wæs þæt gewrixle til, / þæt hie on ba healfa bicgan scoldon / freonda feorum' (1304b–1306b), 'that was not a good bargain, where those on both sides had to pay with the lives of dear ones'. The immediately following lines take the focalisation to Hrothgar, upset at the loss of his chief thane (1307–9), thereby foregrounding the human emotional implications of the monster's rage. This human reaction is as automatic, singular and stereotypical as the monster's, arising from self-preservation and social bonding. Yet, the reader's affective response goes beyond simple emotional contagion, because the monster's attack was presented as emotionally motivated. So, while the audience is undoubtedly cheering for the humans, prior engagement with the emotional reaction of a mother grieving for her dead son invites a cognitive, affective readerly response to the recounted battle that is explicitly presented as being between emotionally charged adversaries.

52 An argument for the hand's being Æschere's, not Grendel's, has been made: See *Klaeber's Beowulf and the Fight at Finnsburg*, 4th edn, ed. R. D. Fulk, Robert E. Bjork and John D. Niles (Toronto, 2008), p. 198.

53 In a similar way, Grendel's motivation for atrocity in the first place was presented as unendurable torment at being excluded from the revelry in the hall: he 'earfoðlice/ þrage geþolode' (lines 86b–7a), 'endured that time of torture'. Later, too, the dragon responds to the theft of a goblet enraged ('abealch', line 2280b) at his loss.

Similarly, when the loyal Wiglaf addresses his companions, who abandoned their leader Beowulf at the crucial moment, he enacts a familiar emotional trajectory and invites a sympathetic response from the audience. He combines sorrow for his dead lord and smug scorn for his cowardly comrades: 'Wiglaf maðelode, Weohstanes sunu; / sec sarigferð seah on unleofe' (2862b–2863), 'Wiglaf spoke, son of Weohstan, sick, sad at heart, looked at the unloved ones'. Like the speaker of *Wulf and Eadwacer*, here grief is likened to physical sickness, an embodied experience that is familiar to medieval and modern readers.

The literary discourses of emotion in this culture not only physicalise distress in this recognisable way, but also create a firm connection between the experience and the expression of emotion (3148b–3155):

Higum unrote
modceare mændon, mondryhtnes cw(e)alm,
swylce giomorgyd (Ge)at(isc) meowle
(æfter Biowulfe b)undenheorde
(sang) sorgcearig, sæ(*id*)e (ge)neah(*he*),
þæt hio hyre (here)g(eon)gas hearde ond(r)ede,
wælfylla wo(r)n, (w)erudes egesan,
hy[n]ðo ond hæf(t)nyd. Heofon rece swealg.

> Sad in spirit, they bewailed their heart-sorrow, the death of their lord. Likewise, a sorrowful Geatish woman, with her hair bound up for Beowulf, sang a mournful song, told earnestly that she sorely dreaded invasions, a multitude of killings, the terror of the army, humiliation and enslavement. Heaven swallowed the smoke.

Here, a new text world, projected through imagination and fear, gives detail and multiple perspectives to the grief: not just for the loss of the leader as a person, but for the implications of this loss to the community. Emotion is embodied, socially distributed and publically demonstrated, so its experience and expression occur together. Heaven is personified in the act of swallowing, embodying human suffering via a vivid and recognisable conceptual metaphor.

Clearly, the narrative context has an impact on the experience and complexity of emotions by the reader or hearer of the text, of which the cumulative construction of a mood of retrospection and even doom in the poem is a product. The representation of emotional human behaviour, as well as human cognitive activity, models and triggers similar activities in the mind of the audience – a form of emotional contagion that invites empathy.

The arousal and experience of textual emotions

These texts, like other medieval poems, portray recognisable human emotions, and in turn invite emotional reactions from readers today. We can distinguish the intense and conflicted emotional texture of *Wulf and Eadwacer* and the mixture of grief and vengeance of *Beowulf* precisely because we are familiar with the way in which emotions function in the management of goals and as a stimulus to action. These texts can arouse intense, involved readings from a modern audience because they model emotional reactions and narratively produce emotional experiences and genuine sensations, notwithstanding their acknowledged remoteness and even their fictionality. This readerly implication, or emotional investment, arises from rich, textured features within discourse that can be analysed in the context of ideas about emotion, consciousness and memory that are still emerging from the Cognitive Sciences.

As the analyses above have shown, a consideration of the emotional basis of the mental processes activated during textual interpretation provides an explanation for cross-cultural intelligibility, as well as the affective capacity of a fictional narrative. Cognitive processes grounded in the emotional life determine both individual experience and also literary structure and response. The broad continuities in human emotional experience over time ensure the similarity of the emotional quality of textual creation and reception. Notwithstanding this consistency over time and across locations, emotions are subject at least in part to cultural impact, so can be studied in the context of cultural history. Given the growing interest in the field of the History of Emotions, and the arguments being developed about the particular impact of modern experience on the emotional life, the time is ripe for the Anglo-Saxonist to contribute to a fuller diachronic examination of the affective potential of literary texts, as well as the capacity of these texts to represent human emotional experience that is both embodied (and therefore biologically consistent) and also subject to the localised shaping power of a specific cultural context.

8

Conclusion

Renewing Old English literary studies through cognitive approaches

The developing cognitive approaches to literature outlined above can be applied to Old English texts with results that are significant both in their own right as interpretive responses to the texts, and also in terms of bringing Anglo-Saxonists into emerging conversations between Literary Studies and Cognitive Science. Frequently, the dialogues that are currently occurring at this intersection treat recent or early-modern literary texts at the farthest remove, so are limited in their diachronic scope and cross-cultural potential. By bringing the earliest texts written in English into the discussion, a fuller picture can emerge of how evolved human cognitive abilities and literary capabilities have functioned over a whole literary tradition. In turn, Anglo-Saxon Literary Studies stands to gain from this convergence a wider range of interpretive frameworks and exciting possibilities to entertain into the future.

The consideration of these topics in Literary Studies (though for the most part not by medievalists) has shown how interested scholars have been in Psychology and literature, and therefore how well poised to embrace the emerging and productive dynamic operating between Cognitive Science and Literary Studies. The chapters above have traced some manifestations of that potential as it relates specifically to Anglo-Saxonists. In particular, the pages above have shown how extensively conceptual metaphor and conceptual blending underpin cognitive approaches to Literary Studies, and how prospectively rich and varied these synthetic thought processes could be.

Chapter 2 has indicated how a cognitive approach to metaphor and the formation and transmission of concepts underpinning those metaphors can account for the cross-cultural intelligibility of ideas, conceptual frames and figurative expressions. It has explored two dominant metaphors recurring in Old English poetry – of the mind as a container and as a wandering entity – in order to trace vernacular understandings of the embodied mind and nuanced uses of metaphorical language in habitual association with maritime concepts. The familiarity, and indeed durability, of these conceptual metaphors shows how deeply conventionalised they are in general, and how they are customised in particular cultural contexts. A specifically cognitive approach allows an explanation of the systematic generation and comprehension of groups of metaphors, through a consideration of how biological hardwiring interacts with cultural factors in textual encounters.

Chapter 3, on Conceptual Blending Theory, likewise has provided an avenue into an informed analysis of the cognitive processes at work in our interpretation of literary texts. Its account of the logistics of meaning-making relies on the idea that the human mind is primed to create new information from the combination of existing pieces of data. This important idea of continual, routine conceptual synthesis during thought processes and acts of communication underpins most if not all cognitive approaches, because the synthetic blending of ideas in the mind is the basis of all imaginative projections, inferential reasoning and sense-making generally. Anglo-Saxonists can make use of this idea of ubiquitous blending in order to trace the components, patterns and synthetic products of vernacular conceptual combinations as they are manifested in literary texts.

Chapter 4, on Text World Theory, has explored cognitive activity below the horizon of consciousness during textual interpretation. It has shown how we conceptualise the proposition of texts, in that we create imaginative scenarios that are built up and updated during the process of interpretation. This chapter has concentrated on the creation and management of mental representations from textual clues, complete with changes of time and location through the creation of sub-worlds. This cognitive approach is particularly useful for literary analysis, as it provides a detailed way of accounting for how we integrate linguistic cues with memories, knowledge, cultural schemas and other information in the act of creating coherence. Like other cognitive approaches, it combines well with other approaches to literary interpretation, by drawing attention to what is happening conceptually during meaning-making, especially the structured creation of systems of mental representations triggered by the language of the text.

Chapter 5, on Cognitive Cultural Studies, has addressed a range of approaches to the mutual impact of culture and cognition. It has focused especially on the idea of mind-reading, or Theory of Mind, an idea from Cognitive Science to account for how we assume, from observed behaviour, the mental states that have produced that behaviour. It has shown how pervasive the use of mind-reading is in Old English poetry, and in turn has challenged the developing orthodoxy in Cognitive Cultural Studies that mind-reading is particularly represented in, and tied up with the development of, the modern novel. Rather, mind-reading and complex systems of nested mental states (he thought that she assumed that he said. . .) underpin Old English poetry. This poetry is cognitively complex, and makes strenuous cognitive demands on its audience, just as modern texts do. As the texts explored above have shown, medieval audiences of poetry were required to hold several levels of intentionality in mind during textual processing, and routinely to understand and use their Theory of Mind, to put themselves into someone else's shoes, and to understand their perspective. Greater historical depth and cultural variability can be brought to the study of the Theory of Mind if texts from the earlier medieval period are brought into the discussions more fully. Furthermore, greater diachronic range can only enhance the opportunity of testing fully a key assumption from Cognitive Cultural

Studies, that cognitive processes are determined in part by embedded cultural practices. This chapter, and this book more broadly, appeal to Anglo-Saxon specialists to use their specific expertise to explore the applicability and potential insights such as these to Old English literature.

Chapters 6 and 7 have considered what Anglo-Saxonists have to offer cognitive approaches and even perhaps to Cognitive Science. Chapter 6, on Autobiographical Memory and the Self, has supported arguments for the fundamentally narrative quality of autobiographical memory by using the deep diachronic depth made available through the analysis of the earliest English poetry. This chapter has also explored the capacity of early poetry to represent subjective states and autobiographical memory, and thereby to show the inherent human tendency towards literary self-conceptualisation and the culturally specific and universal aspects of self-reflection and self-presentation. This study of Anglo-Saxon poetic representations of personal identity (the self), and its reliance on autobiographical memory, has thereby sought to contribute to emerging arguments for the dialogic nature of the self and its indebtedness to transmitted schemas. In particular, it has argued for the role played by literary texts to formulate and in turn to transmit models of the self. Anglo-Saxonists can contribute to the future understanding of textual-cognitive interaction by entering into conversations with Cognitive Science, and by challenging, as this chapter has done, some longstanding assumptions by psychologists about medieval self formation and expression.

Chapter 7, on the history of emotions, has likewise shown how Anglo-Saxon perspectives can be brought into a fruitful combination with cognitive approaches to Literary Studies and with Cognitive Science. It has taken as its focus the idea that Old English poetry relies, as does later literature, on a shared understanding of human emotions registered in the body as well as in the mind. This literature can be analysed to consider how emotions were experienced and represented in remote cultures, a subject of much interest in current scholarly circles. Moreover, this chapter has shown how Old English poetry is able to represent recognisable emotions and to trigger emotional states in modern readers, and how that reaction operates within a system of textual processing. This chapter has approached this topic through literary analysis by deploying ideas from Text World Theory, Conceptual Metaphor Theory and the cultural history of emotions. It has argued for Anglo-Saxonists to take part in shaping the burgeoning field, the history of emotions, and to enrich literary analysis by making use of selected key ideas from Cognitive Science relating to emotion, consciousness and memory.

The chapters above have all considered cross-cultural intelligibility – of the self, memory, emotions, the inner life and their textual representation. They have presented the argument that cognitive approaches share preoccupations sourced from Cognitive Science that have a great deal to offer Anglo-Saxonists: conceptual blending; the universal human ability to imagine someone else's mental perspective; and the mutual impact of the embodied mind and its cultural context. Cognitive readings disclose in the temporally remote literary texts produced in Anglo-Saxon England signs of familiar cognitive processing. Such readings show how medieval vernacular poetic texts, and responses to that verse, are structured by recognisable human cognitive functions. Similarly, the evident reliance in Old English verse on conceptual blending, familiar conceptual metaphors, the creation of text worlds, and on a narrative view of the self, supports the view that culture and cognition interact in consistent ways over huge timespans. This new understanding in turn shows the way forward for a range of new perspectives on both early literature and human psychology, a future that relies on far-reaching and mutually respected interdisciplinarity.

The relevance of literary texts for Cognitive Science is clear: as complex products of, and linguistic entities processed by, the human mind, literary texts entail cognitive processing at a very high level. Because there is growing evidence from multiple directions that cultural context has a role in shaping how that processing works, and a longstanding view that culture is in part a product of human cognition, there is a growing imperative to consider culture and cognition as an interdependent dynamic. Therefore, any worthwhile cognitive ecology – or, the study of cognitive phenomena in context – can and must develop a diachronic dimension. Such an ecology must also address literary texts as the highly worked expressive and creative outputs of the human mind, as well as products of culturally specific contexts. In this way, higher-level cognitive functioning and intermental thought embedded in and communicated via textual representations is available for examination, and a new perspective on cross-cultural intelligibility and the consistency of human cognitive functioning can be approached. Old English poetry can be viewed as the product of mental functioning, as well as a representation of that functioning, and even at times as a discussion of it. As a product of an age prior to the explicit study of Psychology, this verse is of particular interest in its intuitive rather than scholarly appeal to the hearts and minds of its immediate audience, and its durable capacity to engage the minds of remote audiences whose cognitive architecture operates similarly in a different cultural context.

Taking some of these ideas further, it might be argued that the implications for our view of the cognitive structure of poetry generally, and the mental functioning triggered by individual examples in the act of interpretation more specifically, are far reaching. Such considerations would require a great deal of close study from more perspectives than this book has been able to canvas. But hopefully, the demonstrated capacity for Old English poetry to feature cognitively complex characteristics recognised in far more recent texts, and its suitability for the approaches outlined above, makes the case for the need to adjust some key conceptions currently held about cognitive literary structure as a distinctive feature of early modernity or, more precisely, of the novel. Furthermore, the apparent compatibility of many of the approaches outlined above – both with each other and also with existing literary, stylistic and linguistic approaches to texts – suggests the benefit of more robust interaction among disciplines currently exploring cognitive approaches, particularly across the literary/linguistic divide. It is to be hoped that Cognitive Poetics, for instance, will develop closer interactions with Cognitive Cultural Studies in future. Similarly, scholars currently working in the humanities would acquire further range and depth by developing familiarity with cognitive approaches to literature and culture, especially those who make the mind the subject of study. Studies of earlier mental models in general, and Anglo-Saxon psychologies in particular, in future would ideally be aware of and engage with developments in Cognitive Science, rather than treat the history of ideas and emotions without that essential cross-disciplinary perspective. In turn, Cognitive Science can learn from Anglo-Saxon studies.

What Old English studies can do for Cognitive Science

Beyond Literary Studies, the arguments made above about the cognitive structures of Old English texts can be deployed to expand or challenge orthodoxies in Cognitive Science. For instance, a consideration of how Old English poetry can be comprehended by, and even evoke a cognitive or an affective response in, modern audiences provides a much larger diachronic sweep to the question of cross-cultural intelligibility than has been provided to date. Similarly, the historical depth afforded by bringing the oldest English poetry into ongoing discussions in cognitive approaches to literature, and also Cognitive Science, provides an opportunity for far more extensive considerations of cognition as a cultural process than has been achieved in any field thus far. In another instance, the narrative structure of autobiographical memory, and its instantiation in and transmission through literary texts, has a far longer genealogy than is usually considered by psychologists. Bringing medieval literary representations of the self and autobiographical memory into current debates has the capacity to overturn dubious but longstanding orthodoxies in the history of Psychology. Likewise, the emerging field, the History of Emotions, can usefully embrace Old English texts and at the same time engage with specifically cognitive approaches to the determination of literary structure and response, while maintaining a rigorous historical grounding.

Most importantly, the nature of the dynamic interplay between culture and cognition can be explored in a far more nuanced way than has been attempted by scholars working within Cognitive Science. As Pascal Boyer expresses it, one of the valuable contributions of combining the scientific with the humanities mode of scholarship is to explore the 'role of cognition in culture',[1] a huge project that has just begun to be broached in any serious manner. From the other side of the humanities/science divide, F. Elizabeth Hart contends that 'our approaches to literary questions can and should be enriched by an acknowledgment of how the enabling and constraining behaviors of brains and minds contribute to literary experience'.[2]

1 Pascal Boyer, 'Why Evolved Cognition Matters to Understanding Cultural Cognitive Variations', *Interdisciplinary Science Reviews* 35.3–4 (2010), 376–86, at p. 376.

2 F. Elizabeth Hart, 'The View of Where We've Been and Where We'd Like to Go', *College*

A mutual recognition of the enabling power of disciplinary cross-fertilization can only enhance our endeavours to understand the deep mysteries of consciousness, mental processing and the cognitive mechanics of meaning-making, as well as the deep inter-reliance of culture and cognition.

This study contends that this proposed interaction among many fields of inquiry offers considerable explanatory and expansive potential, as well as specific benefits for our own field. The viability of Anglo-Saxon Studies into the future, but also the validity of many lines of inquiry existing within Cognitive Science and solidifying within cognitive approaches to literature, can be made more robust by considering the unlikely synergies produced by bringing the oldest English texts into proximity with the newest brain science. Hopefully, this book will prompt further conversations across that daunting divide.

Literature 33.1 (2006), 225–37, at p. 226, reviewing Peter Stockwell, *Cognitive Poetics*, Joanna Gavins and Gerald Steen, *Cognitive Poetics in Practice*, and Gilles Fauconnier and Mark Turner, *The Way We Think*.

Bibliography

Abbott, H. Porter, *The Cambridge Introduction to Narrative*, 2nd edn (Cambridge, 2008)

Abu-Lughod, Lila, and Catherine A. Lutz, 'Introduction: Emotion, Discourse, and the Politics of Everyday Life', in *Language and the Politics of Emotion*, ed. Lutz and Abu-Lughod (Cambridge, 1990), pp. 1–23

Ackerman, Robert W., 'The Debate of the Body and the Soul and Parochial Christianity', *Speculum* 37.4 (1962), 541–65

Adamopoulos, John, 'Analysis of Interpersonal Structures in Literary Works of Three Historical Periods', *Journal of Cross-Cultural Psychology* 13.2 (1982), 157–68

— and Robert N. Bontempo, 'Diachronic Universals in Interpersonal Structures: Evidence from Literary Sources', *Journal of Cross-Cultural Psychology* 17.2 (1982), 169–89

Adams, Glenn, and Vivan Afi Dzokoto, 'Self and Identity in African Studies', *Self and Identity* 2 (2003), 345–59

Adler, Hans, and Sabine Gross, 'Adjusting the Frame: Comments on Cognitivism and Literature', *Poetics Today* 23.2 (2002), 195–220

Ælfric, 'The Nativity of Our Lord Jesus Christ', in *Lives of the Saint*, ed. Walter W. Skeat, 2 vols (1844; repr. Cambridge, 1999), vol. 1, pp. 10–25

Aertsen, Henk, '*Wulf and Eadwacer*: A Woman's *cri de coeur* – For Whom, For What?', in *Companion to Old English Poetry*, ed. Henk Aertsen and Rolf H. Bremmer Jr (Amsterdam, 1994), pp. 119–44

Allen, Michael J. B., and Daniel G. Calder, *Sources and Analogues of Old English Poetry: The Major Latin Texts in Translation* (Cambridge, 1976)

Anderson, S. J., and M. A. Conway, 'Investigating the Structure of Autobiographical Memories', *Journal of Experimental Psychology: Learning, Memory and Cognition* 19 (1993), 1178–96

Andersen, Susan M., and Lee Ross, 'Self-Knowledge and Social Inference: I. The Impact of Cognitive/Affective and Behavioral Data', *Journal of Personality and Social Psychology* 46.2 (1984), 280–93

Apperly, Ian, *Mindreaders: The Cognitive Basis of 'Theory of Mind'* (Hove and New York, 2011)

Ashmore, Richard D., and Lee Jussim, eds, *Self and Identity: Fundamental Issues* (New York and Oxford, 1997)

Augustine, *Confessions*, trans. Henry Chadwick (Oxford, 1992)

Baker, Peter S., 'The Ambiguity of *Wulf and Eadwacer*', *Studies in Philology* 78.5 (1981), 39–51

Bal, Mieke, *Narratology: Introduction to the Theory of Narrative*, 2nd edn (Toronto, 1997)

Bamberg, Michael, 'Who am I? Narration and its Contribution to Self and Identity', *Theory and Psychology* 21.1 (2011), 3–24

Banaji, M. R., and D. A. Prentice, 'The Self in Social Contexts', *Annual Review of Psychology* 45, ed. L. Porter and M. Rosenzweig (1994), 297–332

Barcelona, Antonio, ed., *Metaphor and Metonymy at the Crossroads: A Cognitive Perspective* (Berlin and New York, 2000)

Barnier, Amanda J., John Sutton, Celia B. Harris, and Robert A. Wilson, 'A Conceptual and Empirical Framework for the Social Distribution of Cognition: The Case of Memory', *Cognitive Systems Research* 9 (2008), 33–51

Baron-Cohen, Simon, *The Maladapted Mind: Classic Readings in Evolutionary Psychology* (Hove, 1997)

—— *Mindblindness: An Essay on Autism and Theory of Mind* (Cambridge, MA, and London, 1995)

—— 'Theories of the Autistic Mind', *Psychologist* 21.2 (2008), 112–16

Barresi, John, and Raymond Martin, 'History as Prologue: Western Theories of the Self', in *The Oxford Handbook of the Self*, ed. Shaun Gallagher (Oxford, 2011), 33–56

Baumeister, Roy F., 'How the Self Became a Problem: A Psychological Review of Historical Research', *Psychological Review* 52.1 (1987), 163–76

—— 'The Self', in *The Handbook of Social Psychology*, 4th edn, ed. Daniel T. Gilbert, Susan T. Fiske and Gardner Lindzey, 2 vols. (New York, 1998), vol. I, pp. 680–740

—— 'The Self and Society: Changes, Problems, and Opportunities', in *Self and Identity: Fundamental Issues*, ed. Richard D. Ashmore and Lee Jussim (New York and Oxford, 1997), pp. 191–217

Bede, *Ecclesiastical History of the English People*, ed. Bertram Colgrave and R. A. B. Mynors (Oxford, 1969, repr. 1992)

Bedford, R., L. Davis and P. Kelly, eds, *Early Modern Autobiography: Theories, Genres, Practices* (Ann Arbor, 2006)

Belzen, J. A., and A. Geels, eds, *Autobiography and the Psychological Study of Religious Lives* (Amsterdam and New York, 2008)

Benton, John F., 'Consciousness of Self and Perceptions of Individuality', in *Renaissance and Renewal in the Twelfth Century*, ed. Robert L. Benson and Giles Constable (Cambridge, MA, 1982), pp. 263–95

Berlin, Gail Ivy, 'Grendel's Advance on Heorot: The Functions of Anticipation', *Proceedings of the PMR Conference* 11 (1986), 19–26

Birchmore, Susan, 'Traditional Exegesis and the Question of Guilt in the Old English *Genesis B*', *Traditio* 41 (1985), 117–44

Bjork, Robert E., 'Speech as Gift in *Beowulf*', *Speculum* 69 (1994), 993–1022

—— 'Digressions and Episodes', in *A Beowulf Handbook*, ed. R. Bjork and J. Niles (Lincoln, NE, 1997), pp. 193–212

Boitani, Piero, and Anna Torti, eds, *The Body and the Soul in Medieval Literature* (Cambridge, 1999)

Bonjour, A., *The Digressions in Beowulf* (Oxford, 1950)

Bortolussi, Marisa, and Peter Dixon, *Psychonarratology: Foundations for the Empirical Study of Literary Response* (New York, 2002)

Bosworth, J., and T. N. Toller, *An Anglo-Saxon Dictionary* (1898); *Supplement* by T. N. Toller (1921), with *Enlarged Addenda and Corrigenda* by A. Campbell (Oxford, 1972)

Boyd, Brian, 'Fiction and Theory of Mind', *Philosophy and Literature* 30 (2006), 571–81

—— *On the Origin of Stories: Evolution, Cognition and Fiction* (Cambridge, MA, and London, 2009)

Boyer, Pascal, 'Why Evolved Cognition Matters to Understanding Cultural Cognitive Variations', *Interdisciplinary Science Reviews* 35.3–4 (2010), 376–86

—— and James V. Wertsch, *Memory in Mind and Culture* (Cambridge, 2009)

Brandt, Line, and Per Aage Brandt, 'Cognitive Poetics and Imagery', *European Journal of English Studies* 9.2 (2005), 117–30

Breithaupt, Fritz, 'How is it Possible to Have Empathy? Four Models', in *Theory of Mind and Literature*, ed. P. Leverage, Howard Mancing, Richard Schweickert and Jennifer Marston William (West Lafayette, 2011), pp. 273–88

Brent, John Justin, 'The Legend of Soul and Body in Medieval England', Ph.D. thesis, State University of New York, Stony Brook, 2000, DAI-A 61/09 (2001)

Brockmeier, J., and R. Harré, 'Narrative: Problems and Promises of an Alternative Paradigm', in *Narrative and Identity: Studies in Autobiography, Self and Culture*, ed. J. Brockmeier and D. Carbaugh (Amsterdam and Philadelphia, 2001), pp. 59–73

Broughton, T. L., ed., *Autobiography*, 4 vols (Abingdon, 2007)

Bruner, Jerome S., *Acts of Meaning* (Cambridge, MA, 1990)

— 'The Narrative Construction of Reality', *Critical Inquiry* 18 (1991), 1–21

— 'The Autobiographical Process', in *The Culture of Autobiography: Constructions of Self-Representation*, ed. R. Folkenflik (Stanford, 1993), pp. 38–56

— 'Self-Making and World-Making', in *Narrative and Identity: Studies in Autobiography, Self and Culture*, ed. J. Brockmeier and D. Carbaugh (Amsterdam and Philadelphia, 2001), pp. 25–37

— *Making Stories: Law, Literature, Life* (New York, 2002)

— and Susan Weisser, 'The Invention of the Self: Autobiography and its Forms', in *Literacy and Orality*, ed. David R. Olson and Nancy Torrance (Cambridge, 1991), pp. 129–47

Busse, W. G., and R. Holtei, '*The Battle of Maldon*: A Historical, Heroic, and Political Poem', *Neophilologus* 65 (1981), 614–21

Butte, George, *I Know that You Know that I Know: Narrating Subjects from Moll Flanders to Marnie* (Columbus, 2004)

Bynam, Carolyn Walker, *Jesus as Mother: Studies in the Spirituality of the High Middle Ages* (Berkeley, 1982)

Carnicelli, Thomas A., ed., *King Alfred's Version of Augustine's Soliloquies* (Cambridge, MA, 1969)

Carroll, Joseph, 'Evolutionary Approaches to Literature and Drama', in *The Oxford Handbook of Evolutionary Psychology*, ed. R. I. M. Dunbar and Louise Barrett (Oxford, 2007), pp. 637–48

— *Literary Darwinism: Evolution, Human Nature and Literature* (London and New York, 2004)

Carroll, Noël, 'On the Ties that Bind: Characters, the Emotions, and Popular Fictions', in *Philosophy and the Interpretation of Pop Culture*, ed. William Irwin and Jorge J. E. Gracia (Lanham, 2007), pp. 89–116

Carruthers, M., *The Book of Memory: A Study of Memory in Medieval Culture* (Cambridge, 1990)

— *The Craft of Thought: Meditation, Rhetoric, and the Making of Images, 400–1200* (Cambridge, 1998)

Carver, C. S., and M. F. Scheier, *Attention and Self-Regulation: A Control Theory Approach to Human Behavior* (New York, 1981)

Cavill, Paul, '"Engel Dryhtnes" in "The Dream of the Rood" 9b Again', *Neuphilologische Mitteilungen* 86 (1985), 287–92

Charteris-Black, Jonathan, *Corpus Approaches to Critical Metaphor Analysis* (Houndmills, 2004)

Chemero, Anthony, *Radical Embodied Cognitive Science* (Cambridge, MA, and London, 2009)

Chiao, Joan Y., ed., *Cultural Neuroscience: Cultural Influences on Brain Function* (Amsterdam, Oxford and New York, 2009)

Clark, Andy, *Being There: Putting Brain, Body, and World Together Again* (Cambridge, MA, 1997)
—— *Supersizing the Mind: Embodiment, Action, and Cognitive Extension* (Oxford, 2011)
—— and David J. Chalmers, 'The Extended Mind', *Analysis* 58 (1998), 10–23
Clarke, Basil, *Mental Disorder in Earlier Britain: Exploratory Studies* (Cardiff, 1975)
Clemoes, Peter, 'Action in *Beowulf* and Our Perception of it', in *Old English Poetry: Essays on Style*, ed. Daniel G. Calder (Berkeley, 1979), pp. 147–68
—— *Interactions of Thought and Language in Old English Poetry* (Cambridge, 1995)
—— '*Mens absentia cogitans* in *The Seafarer* and *The Wanderer*', in *Medieval Literature and Civilization: Studies in Memory of G. N. Garmonsway*, ed. D. A. Pearsall and R. A. Waldron (London, 1969), pp. 62–77
Clover, Carol, 'The Germanic Context of the Unferth Episode', *Speculum* 55 (1980), 444–68
Cole, M., *Cultural Psychology: A Once and Future Discipline* (Cambridge, MA, 1996)
Coleman, J., *Ancient and Medieval Memories: Studies in the Reconstruction of the Past* (Cambridge, 1992)
Connor, Patrick W., 'Religious Poetry', in *A Companion to Anglo-Saxon Literature*, ed. P. Pulsiano and E. Treharne (Oxford, 2001), pp. 251–67
Conway, Martin A., and Christopher W. Pleydell-Pearce, 'The Construction of Autobiographical Memories in the Self-Memory System', *Psychological Review* 107.2 (2000), 261–88
Cook, Patrick, '*Woriað þa winsalo*: The Bonds of Exile in "The Wanderer"', *Neophilologus* 80 (1996), 127–37
Cooper, David L., 'Attractor Dynamics in *Beowulf*', in *Linguistic Attractors: The Cognitive Dynamics of Language Acquisition and Change*, Human Cognitive Processing 2 (Amsterdam and Philadelphia, 1999), pp. 206–41
Coplan, Amy, 'Catching Characters' Emotions: Emotional Contagion Responses to Narrative Fiction Film', *Film Studies* 8 (2006), 26–38
—— 'Empathetic Engagement with Narrative Fictions', *Journal of Aesthetics and Art Criticism* 62.2 (2004), 141–52
Corballis, Michael C., *The Recursive Mind: The Origins of Human Language, Thought, and Civilization* (Princeton, 2011)
Coulson, Seana, *Semantic Leaps: Frame-Shifting and Conceptual Blending in Meaning Construction* (Cambridge, 2001)
—— and Todd Oakley, *Journal of Pragmatics* 37.10 (2005)
Crane, John Kenny, 'To Thwack or Be Thwacked: An Evaluation of Available Translations and Editions of *Beowulf*', *College English* 32 (1970), 321–40
Crane, Mary Thomas, 'Analogy, Metaphor, and the New Science: Cognitive Science and Early Modern Epistemology', in *Introduction to Cognitive Cultural Studies*, ed. Lisa Zunshine (Baltimore, 2010), pp. 103–14
—— '"Fair is Foul": *Macbeth* and Binary Logic', in *The Work of Fiction: Cognition, Culture and Complexity*, ed. Alan Richardson and Ellen Spolsky (Aldershot and Burlington, 2004), pp. 107–26
—— *Shakespeare's Brain: Reading with Cognitive Theory* (Princeton, 2004)
Crisp, Peter, 'Between Extended Metaphor and Allegory: Is Blending Enough?', *Language and Literature* 17.4 (2008), 291–308
Cross, Susan E., 'Culture, the Self, and Identity: Introduction to the Special Issue', *Self and Identity* 2 (2003), 265–8
Culpeper, Jonathan, *Language and Characterisation* (London, 2001)

Currie, Gregory, 'The Paradox of Caring: Fiction and the Philosophy of Mind', in *Emotion and the Arts*, ed. Mette Hjort and Sue Laver (Oxford, 1997), pp. 63–77

Damasio, Antonio R., *Descartes' Error: Emotion, Reason, and the Human Brain* (New York, 1994)

— *The Feeling of What Happens: Body, Emotion and the Making of Consciousness* (London, 1999)

Dancygier, Barbara, and Eve Sweetser, *Mental Spaces in Grammar* (Cambridge, 2005)

— 'The Text and the Story: Levels of Blending in Fictional Narratives', in *Mental Spaces in Discourse and Interaction*, ed. Anders Hougaard and Todd Oakley (Amsterdam and Philadelphia, 2008), pp. 51–78

D'Andrade, R. G., 'Cultural Meaning Systems', in *Culture theory: Essays on Mind, Self, and Emotion*, ed. R. A. Shweder and R. A. LeVine (Cambridge, 1984)

— 'A Folk Model of the Mind', in *Cultural Models in Language and Thought*, ed. D. Holland and N. Quinn (Cambridge, 1987), pp. 112–48

Daniëlli, Sonja, '*Wulf, min Wulf*: An Eclectic Analysis of the Wolf-Man', *Neophilologus* 90 (2006), 135–54

Danziger, Kurt, 'The Historical Formation of Selves', in *Self and Identity: Fundamental Issues*, ed. Richard D. Ashmore and Lee Jussim (New York and Oxford, 1997), pp. 137–59

— *Marking the Mind: A History of Memory* (Cambridge, 2008)

Davidson, Arnold E., 'Interpreting *Wulf and Eadwacer*', *Annuale Mediaevale* 16 (1975), 24–32

Davis, Glenn M., 'Changing Senses in *Genesis B*', *Philological Quarterly* 80 (2001), 113–31

— 'Corporeal Anxiety in *Soul and Body II*', *Philological Quarterly* 87.1–2 (2008), 33–50

Davis, M. H., *Empathy: A Social Psychological Approach* (Boulder, 1994)

DeGregorio, Scott, 'Texts, *Topoi* and the Self: A Reading of Alfredian Spirituality', *Early Medieval Europe* 13 (2005), 79–96

Demos, Virginia E., ed., *Exploring Affect: The Selected Writings of Silvan S. Tomkins* (Cambridge, 1995)

Dennett, D., *Consciousness Explained* (London, 1991)

— 'The Self as a Center of Narrative Gravity', in *Self and Consciousness: Multiple Perspectives*, ed. F. S. Kessel, P. M. Cole and D. L. Johnson (Hillsdale, NJ, 1992), pp. 103–15

Diekstra, F. N. M., '*The Seafarer* 58–66a: The Flight of the Exiled Soul to its Fatherland', *Neophilologus* 55 (1971), 433–46

Diller, Hans-Jürgen, 'The Growth of the English Emotion Lexicon: A First Look at the *Historical Thesaurus of English*', in *Of Dyuersitie & Chaunge of Langage: Essays Presented to Manfred Görlach on the Occasion of his 65th Birthday*, ed. Katja Lenz and Ruth Möhlig, Anglistische Forschungen 308 (Heidelberg, 2002), pp. 103–14

Doane, A. N., ed., *The Saxon Genesis: An Edition of the West Saxon 'Genesis B' and the Old Saxon Vatican 'Genesis'* (Madison, WI, and London, 1991)

Donahue, James J., ''Of this I Can Make No Sense': *Wulf and Eadwacer* and the Destabilization of Meaning', *Medieval Forum* 2004, http://www.sfsu.edu/~medieval/Volume4/Donahue.html

Draaisma, D., *Metaphors of Memory: A History of Ideas about the Mind*, trans. Paul Vincent (Cambridge, 2000)

Dreyfuss, Hubert, and Sean Dorrance Kelly, *All Things Shining: Reading the Western Classics to Find Meaning in a Secular Age* (New York, 2011)

Dunbar, Robin, 'Brain and Behaviour in Primate Evolution', in *Mind the Gap: Tracing the Origins of Human Universals*, ed. Peter M. Kappeler and Joan B. Silk (Heidelberg, Dordrecht, London and New York, 2010), pp. 315–30

—— *Grooming, Gossip, and the Evolution of Language* (Cambridge, MA, 1999)

—— 'The Social Brain Hypothesis', *Evolutionary Anthropology* 6 (1998), 178–90

—— 'The Social Brain: Mind, Language, and Society in Evolutionary Perspective', *Annual Review of Anthropology* 32 (2003), 163–81

Dunning, T. P., and A. J. Bliss, eds, *The Wanderer* (London, 1969)

Dutton, Dennis, *The Art Instinct: Beauty, Pleasure and Human Evolution* (Oxford and New York, 2009)

Eakin, P. J., *Fictions in Autobiography: Studies in the Art of Self Invention* (Princeton, 1985)

—— *Living Autobiographically: How We Create Identity in Narrative* (Ithaca and London, 2008)

Ehrhart, Margaret J., 'Tempter as Teacher: Some Observations on The Vocabulary of the Old English *Genesis B*', *Neophilologus* 59 (1975), 435–46

Eliason, Norman E., 'The Burning of Heorot', *Speculum* 55 (1980), 75–83

Elvin, Mark, 'Between the Earth and Heaven: Conceptions of the Self in China', in *The Category of the Person*, ed. Michael Carrithers, Steven Collins and Steven Lukes (Cambridge, 1985), pp. 156–89

Emmott, Catherine, *Narrative Comprehension: A Discourse Perspective* (Oxford, 1997)

—— 'Reading for Pleasure: A Cognitive Poetic Analysis of "Twists in the Tale" and Other Plot Reversal in Narrative', in *Cognitive Poetics in Practice*, ed. Joanna Gavins and Gerard Steen (London, 2003) pp. 145–59

—— '"Split Selves" in Fiction and in Medical "Life Stories": Cognitive Linguistic Theory and Narrative Practice', in *Cognitive Stylistics: Language and Cognition in Text Analysis*, ed. Elena Semino and Jonathan Culpeper (Amsterdam and Philadelphia, 2002), pp. 153–81

Erll, Astrid, and Ansgar Nünning, eds, *Cultural Memory Studies: An International and Interdisciplinary Handbook* (Berlin and New York, 2008)

Fabiszak, Malgorzata, 'A Semantic Analysis of Emotion Terms in Old English', *Studia Anglica Posnaniensia* 34 (1999), 133–46

Falkenberg, R., W. S. Melion and T. S. Richardson, eds, *Image and Imagination of the Religious Self in Late Medieval and Early Modern Europe* (Turnhout, 2003)

Fauconnier, Gilles, *Mappings in Thought and Language* (Cambridge and New York, 1997)

—— *Mental Spaces: Aspects of Meaning Construction in Natural Language* (Cambridge, MA, 1984; repr. Cambridge, 1994)

—— and Mark Turner, *The Way We Think: Conceptual Blending and the Mind's Hidden Complexities* (New York, 2002)

—— and E. Sweetser, eds, *Spaces, Worlds, and Grammar* (Chicago, 1996)

Feinberg, Todd E., *From Axons to Identity: Neurological Explorations of the Nature of the Self* (New York and London, 2009)

—— 'The Nested Neural Hierarchy of the Self', *Consciousness and Cognition* 20.1 (2011), 4–15

Fernández Martínez, Dolores, 'The Study of Old English Literary Texts from the Perspective of Discourse Analysis: An Overview', *The Grove: Working Papers on English Studies* 8 (2001), 69–79

Finnegan, Robert Emmett, 'Eve and "Vincible Ignorance" in *Genesis B*', *Texas Studies in Literature and Language* 18 (1976), 329–39

Fivush, R., 'Owning Experience: Developing Subjective Perspectives in Autobiographical Narratives', in *The Self in Time :Developmental Perspectives*, ed. C. Moore and K. Lemmon (Mahwah, NJ, 2001)

—— 'The Social Construction of Personal Narratives', *Merrill-Palmer Quarterly* 37 (1991), 59–82

Fludernik, Monika, 'Narratology in the Twenty-First Century: The Cognitive Approach to Narrative', *Publications of the Modern Language Association of America* 125.4 (2010), 924–30

—— 'Natural Narratology and Cognitive Parameters', in *Narrative Theory and the Cognitive Sciences*, ed. David Herman (Stanford, 2003), pp. 243–67

—— *Towards a 'Natural' Narratology* (London, 1996)

—— D. C. Freeman and M. H. Freeman, 'Metaphor and Beyond: An Introduction', *Poetics Today* 20.3 (1999), 383–96

Ford, John C., 'A New Conception of Poetic Formulae Based on Prototype Theory and the Mental Template', *Neuphilologische Mitteilungen* 103 (2002), 205–26

Foucault, M., *Technologies of the Self* (Boston, 1988)

Frank, C. K., and Elise Temple, 'Cultural Effects on the Neural Basis of Theory of Mind', in *Cultural Neuroscience: Cultural Influences on Brain Function*, ed. Joan Y. Chiao (Amsterdam, 2009), pp. 213–23

Frank, Roberta, 'The Ideal of Men Dying with their Lord in *The Battle of Maldon*: Anachronism or *Nouvelle Vague*', in *People and Places in Northern Europe 500–1600: Essays in Honour of Peter Hayes Sawyer*, ed. Niels Lund and Ian Wood (Woodbridge, 1991), pp. 95–106

—— 'Old English *ancor* "anchor": Transformation of a Latin Loanword', in *Germanic Texts and Latin Models: Medieval Reconstructions*, Germania Latina IV, ed. K. E. Olsen, A. Harbus and T. Hofstra (Louvain, 2001), pp. 7–27

Frantzen, Allen J., 'The Body in *Soul and Body I*', *Chaucer Review* 17 (1982), 76–88

Freeman, Margaret H., 'Cognitive Linguistic Approaches to Literary Studies: State of the Art in Cognitive Poetics', in *The Oxford Handbook of Cognitive Linguistics*, ed. Dirk Geeraerts and Hubert Cuyckens (Oxford, 2007), pp. 1175–1202

—— 'Cognitive Mapping in Literary Analysis', *Style* 36.3 (2002), 466–85

—— 'The Fall of the Wall Between Literary Studies and Linguistics: Cognitive Poetics', in *Cognitive Linguistics: Current Applications and Future Perspectives*, ed. Gitte Kristiansen, Michel Achard, Rene Dirven and Francisco J. Ruiz de Mendoza Ibañez (Berlin and New York, 2006), pp. 403–28

—— 'From Substance to Story: Narrative, Identity, and the Reconstruction of the Self', in *Narrative and Identity: Studies in Autobiography, Self and Culture*, ed. J. Brockmeier and D. Carbaugh (Amsterdam and Philadelphia, 2001), pp. 283–98

—— 'The Poem as Complex Blend: Conceptual Mappings of Metaphor in Sylvia Plath's "The Applicant"', *Language and Literature* 14.1 (2005), 25–44

—— 'Poetry and the Scope of Metaphor: Toward a Cognitive Theory of Literature', in *Metaphor and Metonymy at the Crossroads: A Cognitive Perspective*, ed. Antonio Barcelona (Berlin and New York, 2000), pp. 253–82

Frese, Dolores, '*Wulf and Eadwacer*: The Adulterous Woman Reconsidered', in *New Readings on Women in Old English Literature*, ed. Helen Damico and Alexandra Hennessey Olsen (Bloomington, 1990), pp. 273–91

Fulk, R. D., 'Unferth and His Name', *Modern Philology* 85 (1987), 113–27

—— Robert E. Bjork, and John D. Niles, eds, *Klaeber's Beowulf and the Fight at Finnsburg*, 4th edn (Toronto, 2008)

Gallagher, Shaun, *How the Body Shapes the Mind* (Oxford, 2005)
— ed., *The Oxford Handbook of the Self* (Oxford, 2011)
Gameson, Fiona, and Richard Gameson, '*Wulf and Eadwacer, The Wife's Lament*, and the Discovery of the Individual in Old English Verse', in *Studies in Old English Language and Literature: 'Doubt Wisely': Papers in Honour of E. G. Stanley*, ed. M. J. Toswell and E. M. Tyler (London and New York, 1996), pp. 457–74
Ganze, Ronald J., 'Conceptions of the Self in Augustine, King Alfred and Anglo-Saxon England', Ph.D. thesis, University of Oregon, 2004, DAI-A65/11 (2005)
Gavins, Joanna, *Text World Theory: An Introduction* (Edinburgh, 2006)
— '"Too Much Blague?": An Exploration of the Text Worlds of Donald Barthelme's *Snow White*', in *Cognitive Poetics in Practice*, ed. Joanna Gavins and Gerard Steen (London, 2003), pp. 129–44
— and Gerard Steen, eds, *Cognitive Poetics in Practice* (London, 2003)
Geary, Patrick J., *Phantoms of Remembrance: Memory and Oblivion at the End of the First Millennium* (Princeton, 1994)
Geeraerts, Dirk, and Caroline Gevaert, 'Hearts and (Angry) Minds in Old English', in *Culture, Body, and Language: Conceptualizations of Internal Body Organs across Cultures and Languages*, ed. Farzad Sharifian, René Dirven, Ning Yu and Susanne Niemeier (New York and Berlin, 2008), pp. 319–47
— and Stefan Grondelars, 'Looking Back in Anger: Cultural Traditions and Metaphoric Patters', in *Language and the Cognitive Construal of the World*, ed. John R. Taylor and Robert E. MacLaury (Berlin and New York, 1995), pp. 153–79
Gelderen, Elly van, *A History of English Reflexive Pronouns: Person, Self, and Interpretability*, Linguistik Aktuell 39 (Amsterdam and Philadelphia, 2000)
Gergen, Kenneth J., *Relational Being: Beyond Self and Community* (Oxford, 2009)
— 'The Social Construction of Self', in *The Oxford Handbook of the Self*, ed. Shaun Gallagher (Oxford, 2011), pp. 633–53
Gerrig, Richard J., *Experiencing Narrative Worlds: On the Psychological Activities of Reading* (New Haven and London, 1993)
Gevaert, Caroline, 'The Anger is Heat Question: Detecting Cultural Influence on the Conceptualisation of Anger through Diachronic Corpus Analysis', in *Perspectives on Variation: Sociolinguistic, Historical, Comparative*, ed. N. Delbecque, J. van der Auwera and D. Geeraerts (Berlin and New York, 2005), pp. 195–208
Gibbs, Raymond W. Jr, *Embodiment and Cognitive Science* (Cambridge, 2006)
— 'Making Good Psychology out of Blending Theory', *Cognitive Linguistics* 11 (2000), 347–58
— *The Poetics of Mind: Figurative Thought, Language and Understanding* (Cambridge and New York, 1994)
— 'When is Metaphor? The Idea of Understanding in Theories of Metaphor', *Poetics Today* 13.4 (1992), 575–606
— and Gerard Steen, eds, *Metaphor in Cognitive Linguistics* (Amsterdam and Philadelphia, 1999)
Godden, Malcolm, 'Anglo-Saxons on the Mind', in *Learning and Literature in Anglo-Saxon England*, ed. Michael Lapidge and Helmut Gneuss (Cambridge, 1985), pp. 271–98
— and Susan Irvine, eds, *The Old English Boethius: An Edition of the Old English Versions of Boethius' De Consolation Philosophiae* (Oxford, 2009)
Goh, Joshua, and Denise Park, 'Culture Sculpts the Perceptual Brain', in *Cultural Neuroscience: Cultural Influences on Brain Function*, ed. Joan Y. Chiao (Amsterdam, New York and Oxford, 2009), pp. 95–111

Goldie, Peter, 'Imagination and the Distorting Power of Emotion', *Journal of Consciousness Studies* 12.8–10 (2005), 127–39

Gone, Joseph P., Peggy J. Miller and Julian Rappaport, 'Conceptual Self as Normatively Oriented: The Suitability of Past Personal Narrative for the Study of Cultural Identity', *Culture and Psychology* 5.4 (1999), 371–98

Gopnik, Alison, 'Theory of Mind', in *MIT Encyclopaedia of the Cognitive Sciences*, ed. Robert A. Wilson and Frank C. Keil (Cambridge, MA, and London, 1999), pp. 838–40

Gottschall, Jonathan, and David Sloan Wilson, eds, *The Literary Animal* (Evanston, Ill., 2005)

Gordon, Ida, ed., *The Seafarer* (London, 1960; repr. Exeter, 1996)

Gorsuch, Edwin N., 'Emotional Expression in a Manuscript of Bede's *Historia Ecclesiastica*: British Library Cotton Tiberius A XIV', *Semiotica* 83 (1991), 227–49

Grady, J., T. Oakley and S. Coulson, 'Blending and Metaphor', in *Metaphor in Cognitive Linguistics*, ed. R. Gibbs and G. Steen (Philadelphia, 1999), pp. 101–24

Green, Eugene, 'Speech Acts and the Question of Self in Alfred's *Soliloquies*', in *Interdigitations: Essays for Irmengard Rauch*, ed. Gerald F. Carr, Wayne Harbert and Lihua Zhang (Frankfurt am Main, Bern and New York, 1998), pp. 211–18

Greenblatt, Stephen, *Renaissance Self-Fashioning: From More to Shakespeare* (Chicago, 1980)

Greenfield, Stanley B., 'The Authenticating Voice in *Beowulf*', *Anglo-Saxon England* 5 (1976), 51–62

— 'The Old English Elegies', in *Continuations and Beginnings: Studies in Old English Literature*, ed. E. G. Stanley (London, 1966), pp. 142–75

Gutchess, Angela H., and Allie Indeck, 'Cultural Influences on Memory', in *Cultural Neuroscience: Cultural Influences on Brain Function*, ed. Joan Y. Chiao (Amsterdam, New York and Oxford, 2009), pp. 137–50

Hale, Dorothy J., 'Aesthetics and the New Ethics: Theorizing the Novel in the Twenty-First Century', *Publications of the Modern Language Association of America* 124.3 (2009), 896–905

Hallowell, Irving, *Culture and Experience* (Philadelphia, 1955)

Hamilton, Craig, 'Conceptual Integration in Christine de Pizan's *City of Ladies*', in *Cognitive Stylistics: Language and Cognition in Text Analysis*, ed. Elena Semino and Jonathan Culpeper (Amsterdam and Philadelphia, 2002), pp. 1–22

Hammack, P. L., 'Narrative and the Cultural Psychology of Identity', *Personality and Social Psychology Review* 12 (2008), 222–47

Hanna, Robert, and Michelle Maiese, *Embodied Minds in Action* (Oxford, 2009)

Harbus, Antonina, 'A Mind for Hagiography: The Psychology of Resolution in *Andreas*', in *Germanic Texts and Latin Models*, Germania Latina IV, ed. K. E. Olsen, A. Harbus, and T. Hofstra (Leuven, Paris and Stirling, VA, 2001), pp. 127–40

— 'Cognitive Studies of Anglo-Saxon Mentalities', *Parergon* 27.1 (2010), 13–26

— 'Exposure to Life-Writing as an Impact on Autobiographical Memory', *Memory Studies* 20 (2011), 1–15

— *The Life of the Mind in Old English Poetry* (Amsterdam and New York, 2002)

— 'The Maritime Imagination and the Paradoxical Mind in Old English Poetry', *Anglo-Saxon England* 39 (2011), 21–42

— 'The Medieval Concept of the Self in Anglo-Saxon England', *Self and Identity* 1.1 (2002), 77–97

— 'Thinking in Metaphors: Figurative Language and Ideas on the Mind', *Sydney Studies in English* 30 (2004), 3–20

Harbus, Antonina, 'Reading Embodied Consciousness in *Emma*', *Studies in English Literature 1500–1900* 51.4 (2011), 765–82

Hardcastle, Valerie Gray, *Constructing the Self*, Advances in Consciousness Research 73 (Amsterdam, 2008)

Harding, Jennifer, and E. Deidre Pribram, eds, *Emotions: A Cultural Studies Reader* (London, 2009)

Harwood, Robin, *The Survival of the Self* (Aldershot, 1998)

Harré, Rom, *Cognitive Science: A Philosophical Introduction* (London, 2002)

—— and Grant Gillett, *The Discursive Mind* (Thousand Oaks, London and New Delhi, 1994)

Hart, F. Elizabeth, 'Cognitive Linguistics: The Experiential Dynamics of Metaphor', *Mosaic* 28.1 (1995), 1–23

—— 'Embodied Literature: A Cognitive-Poststructuralist Approach to Genre', in *The Work of Fiction: Cognition, Culture and Complexity*, ed. Alan Richardson and Ellen Spolsky (Aldershot and Burlington, 2004), pp. 85–106

—— 'The Epistemology of Cognitive Literary Studies', *Philosophy and Literature* 25 (2001), 314–34

—— 'The View of Where We've Been and Where We'd Like to Go', *College Literature* 33.1 (2006), 225–37

Haser, Verena, *Metaphor, Metonymy and Experientialist Philosophy: Challenging Cognitive Semantics* (Berlin and New York, 2005)

Hassabis, D., D. Kumaran and E. A. Maguire, 'Using Imagination to Understand the Neural Basis of Episodic Memory', *The Journal of Neuroscience* 27.52 (2007), 14365–74

Hauf, Petra, and Friedrich Försterling, *Making Minds: The Shaping of Human Minds through Social Context* (Amsterdam and Philadelphia, 2007)

Head, P. E., *Representation and Design: Tracing a Hermeneutics of Old English Poetry* (New York, 1997)

Healey, A. diP., J. Holland, D. McDougall and I. McDougall, *Dictionary of Old English* (Toronto, 1986–, ongoing)

Herman, David, ed., *Narrative Theory and the Cognitive Sciences* (Stanford, 2003)

—— 'Stories as a Tool for Thinking', in *Narrative Theory and the Cognitive Sciences* (Stanford, 2003), pp. 163–92

—— ed., *The Emergence of Mind: Representations of Consciousness in Narrative Discourse in English* (Lincoln, 2010)

—— and Becky Childs, 'Narrative and Cognition in *Beowulf*', *Style* 37 (2003), 177–203

Herman, Luc, and Bart Vervaeck, *Handbook of Narrative Analysis* (Lincoln and London, 2005)

Hermans, Hubert J. M., 'The Dialogical Self: A Process of Positioning in Space and Time', in *The Oxford Handbook of the Self*, ed. Shaun Gallagher (Oxford, 2011), 654–80

Hernadi, Paul, 'Why is Literature: A Coevolutionary Perspective on Imaginative Worldmaking', *Poetics Today* 23.1 (2002), 21–42

Higgins, E. Tory, 'The "Self-Digest": Self-Knowledge Serving Self-Regulatory Functions', *Journal of Personality and Social Psychology* 71.6 (1996), 1062–1083

Hill, Joyce, 'Methodologies, Mantras, and Paradigms: Research in Early Medieval English Literature', *Journal of English and Germanic Philology* 105 (2006), 87–101

Hill, Thomas D., 'Punishment According to the Joints of the Body in the Old English "Soul and Body II"', *Notes and Queries* 15 (1968), 409–10

—— 'Punishment According to the Joints of the Body, Again', *Notes and Queries* 16 (1969), 246

Hoek, Michelle, 'Violence and Ideological Inversion in the Old English *Soul's Address to the Body*', *Exemplaria* 10 (1998), 271–85

Hogan, Patrick Colm, 'On Being Moved: Cognition and Emotion in Literature and Film', in *An Introduction to Cognitive Cultural Studies*, ed. Lisa Zunshine (Baltimore, 2010), pp. 237–56

Holland, Dorothy, 'Selves as Cultured: As Told by an Anthropologist who Lacks a Soul', in *Self and Identity: Fundamental Issues*, ed. Richard D. Ashmore and Lee Jussim (New York and Oxford, 1997), pp. 160–90

—— and Naomi Quinn, 'Culture and Cognition', in *Cultural Models in Language and Thought*, ed. Dorothy Holland and Naomi Quinn (Cambridge, 1987), pp. 3–40

—— William Lachicotte Jr, Debra Skinner and Carole Cain, *Identity and Agency in Cultural Worlds* (Cambridge, MA, and London, 1998)

Hougaard, Anders, and Todd Oakley, 'Introduction: Mental Spaces and Discourse Analysis', in *Mental Spaces in Discourse and Interaction*, ed. Anders Hougaard and Todd Oakley (Amsterdam and Philadelphia, 2008), pp. 1–26

Howe, Nicholas, *Migration and Mythmaking in Anglo-Saxon England* (New Haven and London, 1989)

Hultin, N., 'The External Soul in 'The Seafarer' and 'The Wanderer', *Folklore* 88 (1977), 39–45

Hutchins, Edwin, *Cognition in the Wild* (Cambridge, MA, 1995)

Hutto, D. D., *Folk Psychological Narratives: The Socio-Cultural Basis of Understanding Reasons* (Cambridge, MA, 2008)

Hutton, Patrick, 'The History of Mentalities: The New Map of Cultural History', *History and Theory* 20 (1981), 237–59

Iacoboni, Marco, *Mirroring People: The New Science of How We Connect with People* (New York, 2008)

Irvine, M., *The Making of Textual Culture: 'Grammatica' and Literary Theory 350–1100* (Cambridge, 1994)

Irving, E., *A Reading of Beowulf* (New Haven, 1968)

Iser, Wolfgang, 'Interaction between Text and Reader', in *The Book History Reader*, ed. David Finkelstein and Alistair McCleery (London and New York, 2002), pp. 291–6

Jackendoff, Ray, *Language, Consciousness and Culture: Essays on Mental Structure* (Cambridge, MA, 2007)

Jackson, Tony E., 'Issues and Problems in the Blending of Cognitive Science, Evolutionary Psychology, and Literary Study', *Poetics Today* 23.1 (2002), 161–79

Jager, Eric, 'Speech and the Chest: Orality or Pectorality', *Speculum* 65 (1990), 845–59

—— 'The Tempter as Rhetoric Teacher: The Fall of Language in the Old English *Genesis B*', *Neophilologus* 72 (1988), 434–48

—— *The Tempter's Voice: Language and the Fall in Medieval Literature* (Ithaca and London, 1993)

—— 'The Word in the "breost": Interiority and the Fall in *Genesis B*', *Neophilologus* 75 (1991), 279–90

Jagger, Holly Elizabeth, 'Body, Text and Self in Old English Verse: A Study of "Beowulfian" and "Cynewulfian" Rhetoric', Ph.D. thesis, University of Toronto, 2002, DAI 63A (2003)

Jambeck, Thomas J., 'The Syntax of Petition in *Beowulf* and *Sir Gawain and The Green Knight*', *Style* 7.1 (1973), 21–9

Jauss, Hans Robert, *Toward an Aesthetic of Reception*, trans. Timothy Bahti, introd. by Paul de Man (Brighton, 1982)

John, Eric, *Reassessing Anglo-Saxon England* (Manchester and New York, 1996)

Johnson, Frank, 'The Western Concept of the Self', in *Culture and Self: Asian and Western Perspectives*, ed. Anthony J. Marsella, George DeVos and Francis L. K. Hsu (New York and London, 1985), pp. 91–138

Johnson, Mark, ed., *Philosophical Perspectives on Metaphor* (Minneapolis, 1981)

Johnson-Laird, P. N., *Mental Models* (Cambridge, 1983)

Josselin, R., and A. Lieblich, 'Reflections on the Narrative Study of Lives', *Narrative Inquiry* 19 (2009), 183–98

Junker, Marie-Odile, 'A Native American View of the "Mind" as Seen in the Lexicon of Cognition in East Cree', *Cognitive Linguistics* 14.2.3 (2003), 167–94

Kashima, Y., E. Kashima, T. Farsides, K. Uichol, F. Strack, L. Werth, Lioba and M. Yuki, 'Culture and Context-Sensitive Self: The Amount and Meaning of Context-Sensitivity of Phenomenal Self Differ across Cultures', *Self and Identity* 3 (2004), 125–41

Kaster, Robert A., *Emotion, Restraint and Community in Ancient Rome* (Oxford and New York, 2005)

Kay, Christian J., 'Metaphors We Have Lived By: Pathways between Old and Modern English', in *Essays on Anglo-Saxon and Related Themes in Memory of Lynne Grundy*, ed. Jane Roberts and Janet Nelson (London, 2000), pp. 273–85

—Jane Roberts, Michael Samuels and Irené Wotherspoon, *Historical Thesaurus of the Oxford English Dictionary: With Additional Material from a 'Thesaurus of Old English'*, 2 vols (Oxford and New York, 2009)

Kecskes, Istvan, 'Editorial: Lexical Merging, Conceptual Blending, and Cultural Crossing', *Intercultural Pragmatics* 1.1 (2004), 1–26

Keen, Suzanne, 'A Theory of Narrative Empathy', *Narrative* 14.3 (2006), 207–36

Kemp, Simon, *Medieval Psychology* (New York, 1990)

Kendall, Calvin B., 'From Sign to Vision: The Ruthwell Cross and *The Dream of the Rood*', in *The Place of the Cross in Anglo-Saxon England*, ed. Catherine E. Karkov, Sarah Larratt Keefer and Karen Louise Jolly, Publications of the Manchester Centre for Anglo-Saxon Studies 4 (Woodbridge, 2006), pp. 129–44

Keysar, Boak, and Sam Glucksberg, 'Metaphor and Communication', *Poetics Today* 13.4 (1992), 633–58

Keysers, Christian, and Valeria Gazzola, 'Unifying Social Cognition', in *Mirror Neuron Systems: The Role of Mirroring Processes in Social Cognition*, ed. Jaime A. Pineda (New York, 2009), pp. 3–37

Kirschner, S., and J. Martin, eds, *The Sociocultural Turn in Psychology: The Contextual Emergence of Mind and Self* (New York, 2010)

Kitayama, S., and H. Markus, eds, *Handbook of Cultural Psychology* (New York, 2007)

Klinck, Anne L., 'Female Characterisation in Old English Poetry and the Growth of Psychological Realism: Genesis B and Christ I', *Neophilologus* 63 (1979), 597–610

— *The Old English Elegies: A Critical Edition and Genre Study* (Montreal and Kingston, 1992)

Knuuttila, Simo, *Emotions in Ancient and Medieval Philosophy* (Oxford, 2004)

Kondo, Dorinne K, *Crafting Selves: Power, Gender and Discourses in a Japanese Workplace* (Chicago, 1990)

Kövecses, Zoltán, *Metaphor: A Practical Introduction* (New York, 2002)

— 'Metaphor and Thought', in *The Cambridge Handbook of Metaphor and Thought*, ed. Raymond W. Gibbs (Cambridge, 2008), pp. 380–96

— *Metaphor in Culture: Universality and Variation* (Cambridge, 2005)

Kramnick, J., 'Against Literary Darwinism', *Critical Inquiry* 37 (2) (2011), 315–47

Krapp, George Philip, and Elliot van Kirk Dobbie, eds, *The Anglo-Saxon Poetic Records*, 6 vols (New York, 1931–42)
Kringelbach, Morton L., 'Emotions, Feelings, and Hedonics in the Human Brain', in *The Emotions: A Cultural Reader*, ed. Helena Wulff (Oxford and New York, 2007), pp. 37–60
Kristánsson, Kristán, *The Self and its Emotions* (Cambridge, 2010)
Lakoff, George, and Mark Johnson, *Metaphors We Live By* (Chicago, 1980)
— and Mark Johnson, *Philosophy in the Flesh: The Embodied Mind and its Challenge to Western Thought* (New York, 1999)
— and Mark Turner, *More Than Cool Reason: A Field Guide to Poetic Metaphor* (Chicago and London, 1989)
Lane, Richard D., Lynn Nadel, John J. B. Allen and Alfred W. Kasniak, 'The Study of Emotion from the Perspective of Cognitive Neuroscience', in *Cognitive Neuroscience of Emotion*, ed. Richard D. Lane and Lynn Nadel (Oxford, 2000), pp. 3–11
Langacker, Ronald W., *Cognitive Grammar: A Basic Introduction* (Oxford, 2008)
Langeslag, P. S., 'Doctrine and Paradigm: Two Functions of the Innovations in *Genesis B*', *Studia Neophilologica* 79 (2007), 113–18
Lapidge, Michael, '*Beowulf* and Perception', *Publications of the British Academy* 111 (2001), 61–97 (Sir Israel Gollancz Memorial Lecture)
— '*Beowulf* and the Psychology of Terror', in *Heroic Poetry in the Anglo-Saxon Period: Studies in Honour of Jess B. Bessinger Jr*, ed. Helen Damico and John Leyerle, Studies in Medieval Culture XXXII (Kalamazoo, 1993), pp. 373–402
Leary, Mark R., Editorial: 'What is the Self? A Plea for Clarity', *Self and Identity* 3.1 (2004), 1–3.
LeDoux, Joseph, *The Emotional Brain: The Mysterious Underpinnings of Emotional Life* (New York, 1998)
— and Michael Rogan, 'Emotion and the Animal Brain', in *MIT Encyclopedia of the Cognitive Sciences*, ed. Robert A. Wilson and Frank C. Keil (Cambridge, MA, 1999), pp. 269–71
Lee, David, *Cognitive Linguistics: An Introduction* (Oxford and New York, 2001)
Leezenberg, Michiel, *Contexts of Metaphor* (Amsterdam, 2001)
Lerer, Seth, 'Grendel's Glove', *English Literary History* 61 (1994), 721–51
Leslie, R. F., ed., *The Wanderer* (Manchester, 1966)
Leverage, Paula, *Memory and Reception: A Cognitive Approach to the Chansons de Geste* (Amdsterdam, 2010)
— 'Is Perceval Autistic? Theory of Mind in the *Conte del Graal*', in *Theory of Mind and Literature*, ed. Paula Leverage, Howard Mancing, Richard Schweickert and Jennifer Marston William (West Lafayette, IN, 2011), pp. 133–51
— Howard Mancing, Richard Schweickert and Jennifer Marston William, eds, *Theory of Mind and Literature* (West Lafayette, IN, 2011)
LeVine, Robert A., 'Afterword by Robert A. LeVine', in *The Emotions: A Cultural Reader*, ed. Helena Wulff (Oxford and New York, 2007), pp. 397–9
Locherbie-Cameron, Margaret A. L., '"Anglo-Saxon Attitudes": the Visual Nature of Some Poetic Narrative Structures', *Parergon* 10.2 (1992), 71–82
Lockett, Leslie, *Anglo-Saxon Psychologies in the Vernacular and Latin Traditions* (Toronto, 2011)
— 'Corporeality in the Psychology of the Anglo-Saxons', Ph.D. thesis, University of Notre Dame, 2004, DAI-A 65/06 (2004)
— 'The Role of Grendel's Arm in Feud, Law, and Narrative Strategy of Beowulf', in *Latin Learning and English Lore: Studies in Anglo-Saxon Literature for Michael*

Lapidge, 2 vols, ed. Katherine O'Brien O'Keeffe and Andy Orchard (Toronto, 2005), vol. 1, pp. 368–88

Losin, E. A. R., M. Dapretto and M Iacoboni, 'Culture in the Mind's Mirror: How Anthropology and Neuroscience can Inform a Model of the Neural Substrate from Cultural Imitative Learning', in *Cultural Neuroscience: Cultural Influences on Brain Function*, ed. Joan Y. Chiao (Amsterdam, New York and Oxford, 2009), pp. 175–90

Low, Soon-Ai, 'The Anglo-Saxon Mind: Metaphor and Common-Sense Psychology in Old English Literature', Ph.D. thesis, University of Toronto, 1998, DAI-A 60/10 (2000)

— 'Approaches to the Old English Vocabulary for "Mind"', *Studia Neophilologica* 73 (2001), 11–22

— 'Mental Cultivation in *Guthlac B*', *Neophilologus* 81 (1997), 625–36

Lucas, Peter J., 'Loyalty and Obedience in the Old English *Genesis* and the Interpolation of *Genesis B* into *Genesis A*', *Neophilologus* 76 (1992), 121–35

Macdonald, Paul S., *History of the Concept of Mind: Speculations about Soul, Mind and Spirit from Homer to Hume* (Aldershot, 2003)

MacMullen, Ramsay, *Feelings in History, Ancient and Modern* (Claremont, 2003)

Malle, Bertram F., *How the Mind Explains Behaviour: Folk Explanations, Meaning, and Social Interaction* (Cambridge, MA, and London, 2004)

Mancing, Howard, 'Sancho Panza's Theory of Mind', in *Theory of Mind and Literature*, ed. Paula Leverage, Howard Mancing, Richard Schweickert and Jennifer Marston William (West Lafayette, IN, 2011), pp. 123–32

Markus, Hazel Rose, and Shinobu Kitayama, 'Culture and the Self: Implications for Cognition, Emotion, and Motivation', *Psychological Review* 989.2 (1991), 224–53

Marsella, A. J., G. DeVos, and F. L. K. Hsu, *Culture and Self* (New York, 1985)

Martin, Jack, Jeff H. Sugarman and Sarah Hickinbottom, *Persons: Understanding Psychological Selfhood and Agency* (New York, 2010)

Martin, Raymond, and John Barresi, *The Rise and Fall of Soul and Self: An Intellectual History of Personal Identity* (New York, 2006)

Marvin, William Perry, 'Heorot, Grendel, and the Ethos of the Kill', *In Geardagum* 24 (2003), 1–39

Mascuch, M., *Origins of the Individualist Self: Autobiography and Self-Identity in England, 1591–1791* (Cambridge, 1997)

Matto, Michael Eugene, 'Containing Minds: Mind, Metaphor, and Cognition in Old English Literature', Ph.D. thesis, New York University, 1999, DAI 59A (1999)

— 'The Old English *Soul and Body I* and *Soul and Body II*: Ending the Rivalry', *In Geardagum* 18 (1997), 39–58

McAdams, D. P., 'The Case for Unity in the (Post)Modern Self: A Modest Proposal', in *Self and Identity: Fundamental Issues*, ed. Richard D. Ashmore and Lee Jussim (New York and Oxford, 1997), pp. 46–78

— *The Redemptive Self: Stories Americans Live By* (Oxford, 2006)

— R. Josselson and A. Lieblich, eds, *Identity and Story: Creating Self in Narrative* (Washington, 2006)

McCarthy, J., *Dennett and Ricoeur on the Narrative Self* (Amherst, NY, 2007)

McCormack, Michael, *Origins of the European Economy Communications and Commerce, A.D. 300–900* (Cambridge, 2001)

McC. Gatch, Milton, 'King Alfred's Version of Augustine's *Soliloquies*: Some Suggestions on its Rationale and Unity', in *Studies in Earlier Old English Prose*, ed. P. Szarmach (Albany, 1986), pp. 17–45

McGillivray, Murray, '*Dream of the Rood* 9–12 and the Christian Liturgy', *Notes and Queries*, n.s. 52 (2005), 1–2

McIlwain, James T., 'Brain and Mind in Anglo-Saxon Medicine', *Viator* 37 (2006), 103–12

McLean, K. C., M. Pasupathi and J. L. Pals, 'Selves Creating Stories Creating Selves: A Process Model of Self-development', *Personality and Social Psychology Review* 11 (2007), 262–78

Menary, R., 'Embodied Narratives', *Journal of Consciousness* 15 (2008), 63–84

—— *The Extended Mind* (Cambridge, MA, and London, 2010)

Meskin, Aaron, and Johathan M. Weinberg, 'Emotions, Fiction, and Cognitive Architecture', *British Journal of Aesthetics* 43.1 (2003), 18–34

Miller, Thomas, ed., *The Old English Version of Bede's Ecclesiastical History of the English People*, Early English Text Society, os 95, 96, 110, 111 (Rochester and London, 1890–8; repr. Cambridge, 2003)

Mintz, Susannah B., 'Words Devilish and Divine: Eve as Speaker in *Genesis B*', *Neophilologus* 81 (1997), 609–23

Mitchell, B., and F. C. Robinson, *A Guide to Old English*, 5th ed. (Oxford, 1992)

—— '*Dream of the Rood* Repunctuated', in *Words and Works: Studies in Medieval English Literature in Honor of Fred C. Robinson*, ed. Nicholas Howe and Peter S. Baker (Toronto, 1998), 143–57

Mize, Britt, 'The Representation of the Mind as an Enclosure in Old English Poetry', *Anglo-Saxon England* 35 (2006), 57–90

—— 'Manipulations of the Mind-as-Container Motif in *Beowulf, Homiletic Fragment II*, and Alfred's *Metrical Epilogue to the Pastoral Care*', *Journal of English and Germanic Philology* 107 (2008), 25–56

Moffat, Douglas, 'The MS Transmission of the OE *Soul and Body*', *Medium Ævum* 52 (1983), 300–2

—— *The Old English Soul and Body* (Wolfeboro, NH, 1990)

Morillas, José M. Martín, 'The Concept of Self: Some Cognitive-Cultural Considerations Concerning its Categorization and Expression in Spanish and English', *Language Design* 2 (1999), 1–21

Morris, Colin, *The Discovery of the Individual, 1050–1200* (Toronto, 1972)

Morse, David Wayne, 'A Natural Progression: Cognitive Metaphor as a Structuring Principle in Old English Poetry', Ph.D. thesis, University of Southern California, 1994, DAI 55A (1994)

Nagucka, Ruta, 'Cognitive Approaches for the Understanding of Old English Loanwords', in *Text and Language in Medieval English Prose: A Festschrift for Tadao Kubouchi*, ed. Akio Oizumi, Jacek Fisiak and John Scahill, Studies in English Medieval Language and Literature 12 (Frankfurt am Main, 2005), pp. 163–76

Nelson, K., 'Narrative and the Emergence of a Consciousness of Self', in *Narrative and Consciousness: Literature, Psychology and the Brain*, ed. G. D. Fireman, Ted E. McVay, Jr, and Owen J. Flanagan (Oxford and New York, 2003), pp. 17–36

—— 'Narrative and Self, Myth and Memory: Emergence of the Cultural Self', in *Autobiographical Memory and the Construction of a Narrative Self: Developmental and Cultural Perspectives*, ed. R. Fivush and C. A. Haden (Mahwah, NJ, 2003), pp. 3–28

Nerlich, Brigitte and David C. Clarke, 'Blending the Past and the Present: Conceptual and Linguistic Integration 1800–2000', in *Metaphor and Metonymy in Comparison and Contrast*, ed. René Dirven and Ralf Pörings (Berlin and New York, 2003), pp. 555–93

Neville, Jennifer, 'Fostering the Cuckoo: Exeter Book *Riddle 9*', *Review of English Studies* 58/236 (2007), 430–46

Nichols, Shaun and Stephen Stich, *Mind Reading: An Integrated Account of Pretence, Self-Awareness and Understanding Other Minds* (Oxford, 2003)

Nicholson, Simon, 'The Expression of Emotional Distress in Old English Prose and Verse', *Culture, Medicine and Psychiatry* 19 (1995), 327–38

Niles, John D., *Beowulf: The Poem and its Tradition* (Cambridge, MA, 1983)

—— *English Heroic Poetry and the Social Life of Texts* (Turnhout, 2007)

—— *Old English Enigmatic Poems and the Play of the Texts* (Turnhout, 2006)

Nisbett, Richard E., *The Geography of Thought: How Asians and Westerners Think Differently . . . and Why* (New York, 2003)

Nussbaum, Martha C., *Upheavals of Thought: The Intelligence of Emotions* (Cambridge, 2001)

Ó Carragáin, Éamonn, *Ritual and the Rood: Liturgical Images and the Old English Poems of the 'Dream of the Rood' Tradition*, British Library Studies in Medieval Culture (London, 2005)

O'Neill, Pamela, *'A Pillar Curiously Engraven, With Some Inscription Upon It': What is the Ruthwell Cross?* British Archaeological Reports, British Series 397 (Oxford, 2005)

Oakley, Ted, and Seana Coulson, 'Connecting the Dots: Mental Spaces and Metaphoric Language in Discourse', in *Mental Spaces in Discourse and Interaction*, ed. Anders Hougaard and Ted Oakley (Amsterdam and Philadelphia, 2008), pp. 27–50

Oatley, Keith, 'A Taxonomy of the Emotions of Literary Response and a Theory of Identification in Fictional Narrative', *Poetics* 23 (1994), 53–74.

—— 'Emotions', in *MIT Encyclopedia of the Cognitive Sciences*, ed. Robert A. Wilson and Frank C. Keil (Cambridge, MA, 1999), pp. 273–5

—— *Emotions: A Brief History* (Malden, MA, 2004)

—— 'Simulation of Substance and Shadow: Inner Emotions and Outer Behavior in Shakespeare's Psychology of Character', *College English* 33.1 (2006), 15–33

—— 'Theory of Mind and Theory of Minds in Literature', in *Theory of Mind and Literature*, ed. Paula Leverage, Howard Mancing, Richard Schweickert and Jennifer Marston William (West Lafayette, IN, 2011), pp. 13–26

—— and P. N. Johnson-Laird, 'Towards a Cognitive Theory of Emotions', *Cognition and Emotion* 1 (1987), 29–50

—— Dacher Keltner and Jennifer M. Jenkins, *Understanding Emotions*, 2nd edn (Oxford, 2006)

Oberman, Lindsay M., and V. S. Ramachandran, 'Reflections on the Mirror Neuron System: Their Evolutionary Functions Beyond Motor Representation', *Mirror Neuron Systems: The Role of Mirroring Processes in Social Cognition*, ed. Jaime A. Pineda (New York, 2009), pp. 39–59

Ogura, Michiko, 'Him self, him selfe, him selfa: A Reflexive Pronoun + Uninflected or Nominative Self', *Studia Neophilologica* 60 (1988), 149–57

—— 'Old and Middle English Verbs of Emotion', *Poetica* 66 (2006), 53–72

—— 'Verbs of Emotion with Reflexive Constructions', in *Lexis and Texts in Early English: Studies Presented to Jane Roberts*, ed. Christian Kay and Louise Sylvester (Amsterdam and Atlanta, 2001), pp. 203–12

—— 'Words of EMOTION in Old and Middle English', in *A Changing World of Words: Studies in English Historical Lexicography, Lexicology and Semantics*, ed. Javier E. Díaz Vera, Costerus n.s. 141 (Amsterdam and Atlanta, 2002), pp. 484–99

Ogura, Michiko, 'Words of Emotion in Old and Middle English Translations of Boethius's *De Consolatione Philosophiae*', in *Text and Language in Medieval English Prose: A Festschrift for Tadao Kubouchi*, ed. Akio Oizumi, Jacek Fisiak and John Scahill, Studies in English Medieval Language and Literature 12 (Frankfurt am Main, 2005), pp. 183–206

Olney, J., *Metaphors of Self: The Meaning of Autobiography* (Princeton, 1972)

Olsen, Karin E., 'Animated Ships in Old English and Old Norse Poetry', in *Animals and the Symbolic in Mediaeval Art and Literature*, ed. L. A. J. R. Houwen, Mediaevalia Groningana 20 (Groningen, 1997), pp. 53–66

Orton, P. R., 'An Approach to *Wulf and Eadwacer*', *Proceedings of the Royal Irish Academy* 85C (1985), 225–58

—— 'The OE "Soul and Body": A Further Examination', *Medium Ævum* 48 (1979), 173–97

—— '*The Seafarer* 58–62a', *Neophilologus* 66 (1982), 450–9

—— 'Spouting Poetry: Cognitive Metaphor and Conceptual Blending in the Old Norse Myth of the Poetic Mead', in *Constructing Nations, Reconstructing Myth: Essays in Honour of T. A. Shippey* , ed. Andrew Wawn, Graham Johnson and John Walter (Turnhout, 2007), pp. 277–300

Osborne, Marijane, '*Dogode* in *Wulf and Eadwacer* and King Alfred's Hunting Metaphors', *American Notes and Queries* 13 (2000), 3–9

—— 'The Text and Context of *Wulf and Eadwacer*', in *The Old English Elegies: New Essays in Criticism and Research*, ed. Martin Green (London and Toronto, 1983), pp. 174–89

Osmond, Rosalie, *Mutual Accusation: Seventeenth-Century Body and Soul Dialogues in their Literary and Theological Context* (Toronto, 1990)

Overing, Gillian R., 'On Reading Eve: *Genesis B* and the Readers' Desire', in *Speaking Two Languages: Traditional Disciplines and Contemporary Theory in Medieval Studies*, ed. Allen J. Frantzen (Albany, 1991), pp. 35–63

Packard, Noel, and Christopher Chen, 'From Medieval Mnemonics to a Social Construction of Memory: Thoughts on Some Early European Conceptualizations of Memory, Morality, and Consciousness', *American Behavioral Scientist* 48.10 (2005), 1297–319

Palmer, Alan, *Fictional Minds* (Lincoln, NE, and London, 2004)

—— *Social Minds in the Novel* (Columbus, 2010)

—— 'Storyworlds and Groups', in *An Introduction to Cognitive Cultural Studies*, ed. Lisa Zunshine (Baltimore, 2010), pp. 176–92

Pandit, Lalita, 'Emotion, Perception and Anagnorisis in *The Comedy of Errors*: A Cognitive Perspective', *College Literature* 33.1 (2006), 94–126

—— and Patrick Colm Hogan, 'Introduction: Morsels and Modules: On Embodying Cognition in Shakespeare's Plays', *College English* 33.1 (2006), 1–13

Paprotté, Wolf, and René Dirven, eds, *The Ubiquity of Metaphor: Metaphor in Language and Thought* (Amsterdam and Philadelphia, 1985)

Parish, Steven M., 'The Sacred Mind: Newar Cultural Representations of Mental Life and the Production of Moral Consciousness', *Ethos* 19.3 (1991), 313–51

Parks, Ward, 'Prey Tell: How Heroes Perceive Monsters in *Beowulf*', *Journal of English and Germanic Philology* 92 (1993), 1–16

Pasnau, Robert, *Theories of Cognition in the Later Middle Ages* (Cambridge, 1997)

Phillips, Michael Joseph, 'Heart, Mind, and Soul in Old English: A Semantic Study', Ph.D. thesis, University of Illinois, 1985, DAI-A 46/07 (1986)

Phillips, Natalie, 'Distraction as Liveliness of Mind: Cognitive Approaches to Characterization in Jane Austen', in *Theory of Mind and Literature*, ed. Paula

Leverage, Howard Mancing, Richard Schweickert and Jennifer Marston William (West Lafayette, IN, 2011), pp. 105–22
Pineda, Jaime A., ed., *Mirror Neuron Systems: The Role of Mirroring Processes in Social Cognition* (New York, 2009)
Polomé, Edgar C., 'Some Comments on the Vocabulary of Emotion in Germanic', in *Interdigitations: Essays for Irmengard Rauch*, ed. Gerald F. Carr, Wayne Harbert and Lihua Zhang (Bern, Frankfurt am Main and New York, 1998), pp. 129–40
Porter, R., ed., *Rewriting the Self: Histories from the Renaissance to the Present* (London, 1997)
Premack, David, and Guy Woodruff, 'Does the Chimpanzee have a Theory of Mind?', *Behavioural and Brain Sciences* 1 (1978), 515–26
Prinz, Jesse, 'Are Emotions Feeling?', *Journal of Consciousness Studies* 12.8–10 (2005), 9–25
Quinn, Naomi, 'The Cultural Basis of Metaphor', in *Beyond Metaphor: The Theory of Tropes in Anthropology*, ed. James W. Fernandez (Stanford, 1991), pp. 56–93
Rachlin, Howard, *The Psychology and Economics of Self Control* (Cambridge, MA, 2000)
Radding, Charles M., *A World Made by Men: Cognition and Society, 400–1200* (Chapel Hill and London, 1985)
—— 'Evolution of Medieval Mentalities: A Cognitive-Structural Approach', *American Historical Review* 83 (1978), 577–97
Radvansky, G. A., D. E. Copeland and R. A. Zwaan, 'A Novel Study: Investigating the Structure of Narrative and Autobiographical Memories', *Memory* 13 (2005), 796–814
Raggatt, Peter T. F., 'Multiplicity and Conflict in the Dialogical Self: A Life-Narrative Approach', in *Identity and Story: Creating Self in Narrative*, ed. D. P. McAdams, R. Josselson and A. Lieblich (Washington, 2006), pp. 15–35
Rathbone, Clare, Martin A. Conway and Chris J. A. Moulin, 'Remembering and Imagining: The Role of the Self', *Consciousness and Cognition* 20.4 (2011), 1175–82
Reese, E., and R. Fivush, 'The Development of Collective Remembering', *Memory* 16 (2008), 201–12
Renoir, Alain, 'Eve's I. Q. Rating: Two Sexist Views of *Genesis B*', in *New Readings on Women in Old English Literature*, ed. Helen Damico and Alexandra Hennessey Olsen (Bloomington and Indianapolis, 1990), pp. 262–72
—— 'Point of View and Design for Terror in *Beowulf*', *Neuphilologische Mitteilungen* 63 (1962), 154–67
—— 'The Self-Deception of Temptation: Boethian Psychology in *Genesis B*', in *Old English Poetry: Fifteen Essays*, ed. Robert P. Creed (Providence, 1967), pp. 47–67
—— '*Wulf and Eadwacer*: A Non-Interpretation', in *Medieval and Linguistic Studies in Honor of Francis Peabody Magoun Jr*, ed. J. B. Bessinger Jr and R. P. Creed (New York, 1965), pp. 147–63
Richardson, Alan, 'Facial Expression Theory from Romanticism to the Present', in *An Introduction to Cognitive Cultural Studies*, ed. Lisa Zunshine (Baltimore, 2010), pp. 65–83
—— 'Studies in Literature and Cognition: A Field Map', in *The Work of Fiction: Cognition, Culture and Complexity*, ed. Alan Richardson and Ellen Spolsky (Aldershot and Burlington, 2004), pp. 1–29
—— and Ellen Spolsky, eds, *The Work of Fiction: Cognition, Culture and Complexity* (Aldershot and Burlington, 2004)

Richardson, Alan, and Ellen Spolsky, 'Preface', in *The Work of Fiction: Cognition, Culture and Complexity*, ed. Alan Richardson and Ellen Spolsky (Aldershot and Burlington, 2004), pp. viii-ix

— and Francis F. Steen, 'Literature and the Cognitive Revolution: An Introduction', *Poetics Today* 23.1 (2002), 1–8

— and Francis F. Steen, 'Reframing the Adjustment: A Response to Adler and Gross', *Poetics Today* 24.2 (2003), 151–9

Richardson, John D., and Destin N. Stewart, 'Medieval Confession Practices and the Emergence of Modern Psychotherapy', *Mental Health, Religion and Culture* 12.5 (2009), 473–84

Ritchie, David L., *Content and Connection in Metaphor* (Houndmills, 2006)

Rizzolatti, Giacomo, and Corrado Sinigaglia, *Mirrors in the Brain – How Our Minds Share Actions and Emotions*, trans. Frances Anderson (Oxford, 2008)

Robbins, Philip, and Murat Ayded, eds, *The Cambridge Handbook of Situated Cognition* (Cambridge, 2009)

Roberts, Jane, 'Old English *un-* "very" and Unferth', *English Studies* 61 (1980), 289–92

Robinson, Fred C., *Beowulf and the Appositive Style* (Knoxville, 1985)

— 'Retrospection in Old English and Other Germanic Literatures', *The Grove* 8 (2001), 255–76

— *The Tomb of Beowulf and Other Essays on Old English* (Oxford, 1993)

Robinson, Jenefer, *Deeper than Reason: Emotion and its Role in Literature, Music, and Art* (Oxford, 2005)

Roediger, H. L., F. M. Zaromb and A. C. Butler, 'The Role of Repeated Retrieval in Shaping Collective Memory', in *Memory in Mind and Culture*, ed. P. Boyer and J. V. Wertsch (Cambridge, 2009), pp. 138–71

Ronen, Ruth, *Possible Worlds in Literary Theory* (Cambridge, 2004)

Rosaldo, Michelle Z., 'Toward an Anthropology of Self and Feeling', in *Culture Theory: Essays on Mind, Self, and Emotion*, ed. Richard A. Shweder and Robert A. LeVine (Cambridge and New York, 1984), pp. 136–57

Rosenberg, Seymour, 'Multiplicity of Selves', in *Self and Identity: Fundamental Issues*, ed. Richard D. Ashmore and Lee Jussim (New York and Oxford, 1997), pp. 23–45

Rosenwein, Barbara H., *Emotional Communities in the Middle Ages* (Ithaca and London, 2006)

Rowley, Sharon M., *The Old English Version of Bede's 'Historia Ecclesiastica'*, Anglo-Saxon Studies 16 (Cambridge, 2011)

Ryan, Marie-Louise, *Possible Worlds, Artificial Intelligence, and Narrative Theory* (Bloomington, 1992)

Ryner, Bradley D., 'Exchanging Battle: Subjective and Objective Conflicts in *The Battle of Maldon*', *English Studies* 87.3 (2006), 266–76

Salmon, V., '"The Wanderer', 'The Seafarer' and the Old English Conception of the Soul', *Modern Language Review* 55 (1960), 1–10

Salvador, Mercedes, 'The Key to the Body: Unlocking Riddles 42–46', in *Naked Before God: Uncovering the Body in Anglo-Saxon England*, ed. B. C. Withers and J. Wilcox, Medieval European Studies 3 (Morgantown, 2005), pp. 60–96

Sandner, David, 'Tracking Grendel: The Uncanny in *Beowulf*', *Extrapolation* 40 (1999), 162–76

Sauer, Hans, 'Ælfric and Emotion', *Poetica* 66 (2006), 37–52

Schacter, D. L., D. R. Addis and R. L. Buckner, 'Remembering the Past to Imagine the Future: The Prospective Brain', *Nature Reviews Neuroscience* 8 (2007), 657–61

Schank, Roger and Robert Abelson, *Scripts, Plans, Goals and Understanding: An Inquiry into Human Knowledge Structures* (Hillsdale, NJ, 1977)
Schechtman, Marya, 'The Narrative Self', in *The Oxford Handbook of the Self*, ed. Shaun Gallagher (Oxford, 2011), pp. 394–416
Scheck, Helene, 'Seductive Voices: Rethinking Female Subjectivities in *The Wife's Lament* and *Wulf and Eadwacer*', *Literature Compass* 5.2 (2008), 220–7
Sclater, S. D., D. W. Jones, H. Price and C. Yates, eds, *Emotion: New Psychosocial Perspectives* (London, 2009)
Sedgefield, John, ed., *King Alfred's Old English Version of Boethius, De Consolatione Philosophiae* (Darmstadt, 1968)
Sedgefield, W. J., 'Old English Notes', *Modern Language Review* 26 (1931), 74–5
Seigel, Jerrold, *The Idea of the Self: Thought and Experience in Western Europe since the Seventeenth Century* (Cambridge, 2005)
Seltzer, John L, '*The Wanderer* and the Meditative Tradition', *Studies in Philology* 80 (1983), 227–37
Semino, Elena, 'A Cognitive Stylistic Approach to Mind Style in Narrative Fiction', in *Cognitive Stylistics: Language and Cognition in Text Analysis*, ed. Elena Semino and Jonathan Culpeper (Amsterdam and Philadelphia, 2002), pp. 95–122
—— 'Blending and Characters' Mental Functioning in Virginia Woolf's "Lappin and Lapinova"', *Language and Literature* 15 (2006), 55–72
—— *Language and World Creation in Poems and Other Texts* (London, 1997)
—— *Metaphor in Discourse* (Cambridge, 2008)
—— 'Possible Worlds and Mental Spaces in Ernest Hemingway's "A Very Short Story"', in *Cognitive Poetics in Practice*, ed. Joanna Gavins and Gerard Steen, *Cognitive Poetics in Practice* (London, 2003), pp. 83–98
—— and Jonathan Culpeper, eds, *Cognitive Stylistics: Language and Cognition in Text Analysis* (Amsterdam and Philadelphia, 2002)
—— and Jonathan Culpeper, 'Foreword', in *Cognitive Stylistics: Language and Cognition in Text Analysis*, ed. Elena Semino and Jonathan Culpeper (Amsterdam and Philadelphia, 2002), pp. ix-xvi
Sharpe, R. A., *Making the Human Mind* (London, 1990)
Shen, Yeshayahu, 'Cognitive Aspects of Metaphor Comprehension: An Introduction', *Poetics Today* 13.4 (1992) 567–74
Shweder, Richard A., 'Properties of Culture: An Ethnographic View', in *Culture Theory: Essays on Mind, Self, and Emotion*, ed. Richard A. Shweder and Robert A. LeVine (Cambridge and New York, 1984), pp. 67–87
—— and Edmund J. Bourne, 'Does the Concept of the Person Vary Cross-Culturally?', in *Culture Theory: Essays on Mind, Self, and Emotion*, ed. Richard A. Shweder and Robert A. LeVine (Cambridge and New York, 1984), pp. 158–99
—— and Robert A. LeVine, eds, *Culture Theory: Essays on Mind, Self, and Emotion* (Cambridge and New York, 1984)
Siebert, Eve E., 'Body and Soul Poems in Old and Middle English', Ph.D. thesis, Saint Louis University, 2008. DAI-A 70/03 (2009)
Šileikytė, Rūta, '*In Search of the Inner Mind*: Old English *gescead* and Other Lexemes for Human Cognition in King Alfred's Boethius', *Kalbotyra* 54.3 (2004), 94–102
Slingerland, Edward, 'Conceptions of the Self in the Zhuangzi: Conceptual Metaphor Analysis and Comparative Thought', *Philosophy East and West* 54.3 (2004), 322–4
—— *What Science Offers the Humanities: Integrating Body and Culture* (Cambridge, 2008)

Smith, Peter B., and Michael Harris Bond, *Social Psychology across Cultures: Analysis and Perspectives* (New York, 1993)
Sorabji, Richard, *Self: Ancient and Modern Insights about Individuality, Life, and Death* (Chicago, 2006)
Spearing, A. C., *Textual Subjectivity: The Encoding of Subjectivity in Medieval Narratives and Lyrics* (Oxford, 2005)
Spence, Sarah, *Texts and the Self in the Twelfth Century* (Cambridge, 1996)
Spiro, Melford E., 'Is the Western Conception of the Self "Peculiar" within the Context of World Cultures?', *Ethos* 21.2 (1993), 107–53
—— 'Some Reflections on Cultural Determinism and Relativism with Special Reference to Emotion and Reason', in *Culture Theory: Essays on Mind, Self, and Emotion*, ed. Richard A. Shweder and Robert A. LeVine (Cambridge and New York, 1984), pp. 323–46
Spolsky, Ellen, 'Cognitive Literary Historicism: A Response to Adler and Gross', *Poetics Today* 24.2 (2003), 161–83
—— *Gaps in Nature: Literary Interpretation and the Modular Mind* (Albany, 1993)
—— 'Making "Quite Anew": Brain Modularity and Creativity', in *An Introduction to Cognitive Cultural Studies*, ed. Lisa Zunshine (Baltimore, 2010), pp. 84–102
—— *Satisfying Skepticism: Embodied Knowledge in the Early Modern World* (Aldershot, 2001)
—— 'Why and How to Take the Fruit and Leave the Chaff', *SubStance* 30 (2001), 177–98
—— 'Women's Work is Chastity: Lucretia, *Cymbeline*, and Cognitive Impenetrability', in *The Work of Fiction: Cognition, Culture and Complexity*, ed. Alan Richardson and Ellen Spolsky (Aldershot and Burlington, 2004), pp. 51–84
Stanley, Eric, 'Old English Poetic Diction and the Interpretation of *The Wanderer, The Seafarer*, and *The Penitent's Prayer*', *Anglia* 73 (1955), 413–66
—— 'Wulf, My Wolf!', in *Old English and New: Studies in Language and Linguistics in Honor of Frederic G. Cassidy*', ed. Joan H. Hall, Nick Doane, and Dick Ringler (New York, 1992), pp. 46–62
Steen, Gerard, 'Analyzing Metaphor in Literature: With Examples from William Wordsworth's "I Wandered Lonely as a Cloud"', *Poetics Today* 20.3 (1999), 499–522
—— *Understanding Metaphor in Literature: An Empirical Approach* (London and New York, 1994)
Stewart Jr, Thomas, W., 'The Mind and Spirit of Old English *mōd* and *fer(h)ð*: The Interaction of Metrics and Compounding', *Ohio State University Working Papers in Linguistics* 52 (1999), 51–62
Stock, B., *Augustine the Reader: Meditation, Self-Knowledge, and the Ethics of Interpretation* (Cambridge, MA, 1996)
Stockwell, Peter, *Cognitive Poetics: An Introduction* (London, 2002)
—— *Texture: A Cognitive Aesthetics of Reading* (Edinburgh, 2009)
Suddendorf, T., and M. C. Corballis, 'The Evolution of Foresight: What is Mental Time Travel, and is it Unique to Humans?', *Behavioral and Brain Sciences* 30.3 (2007), 299–351
Sutton, John, 'Body, Mind, and Order: Local Memory and the Control of Mental Representations in Medieval and Renaissance Sciences of Self', in *1543 and All That: Word and Image in the Proto-Scientific Revolution*, ed. G. Freeland and A. Corones (Dortrecht, 2000), pp. 117–50
—— 'Exograms and Interdisciplinarity: History, the Extended Mind and the Civilizing Process', in *The Extended Mind*, ed. Richard Menary (Cambridge, MA, and London, 2010), pp. 189–225

Sutton, John, *Philosophy and Memory Traces: Descartes to Connectionism* (Cambridge, 1998)
Swanton, Michael, '*The Battle of Maldon*: A Literary Caveat', *Journal of English and Germanic Philology* 67.3 (1968), 441–50
Sweetser, Eve, and Gilles Fauconnier, 'Cognitive Links and Domains: Basic Aspects of Mental Space Theory', in *Space, Worlds, and Grammar*, ed. Gilles Fauconnier and Eve Sweetser (Chicago and London, 1996), pp. 1–28
Tasioulas, J. A., 'The Mother's Lament: "Wulf and Eadwacer" Reconsidered', *Medium Ævum* 65 (1996), 1–18
Taverniers, Miriam, *Metaphor and Metaphorology: A Selective Genealogy of Philosophical and Linguistic Conceptions of Metaphor from Aristotle to the 1990s* (Gent, 2002)
Taylor, Charles, *Sources of the Self: The Making of Modern Identity* (Cambridge, 1989)
Thorne, A., 'Personal Memory Telling and Personality Development', *Personality and Social Psychology Review* 4 (2000), 45–56
— and K. C. McLean, 'Telling Traumatic Events in Adolescence: A Study of Master Narrative Positioning', in *Autobiographical Memory and the Construction of a Narrative Self: Developmental and Cultural Perspectives* (Mahwah, NJ, 2003), pp. 169–86
Timmer, B. J., 'The Elegiac Mood in Old English Poetry', *English Studies* 24 (1942), 33–44
Tomasello, Michael, *The Cultural Origins of Human Cognition* (Cambridge, MA, 1999)
— 'The Gap is Social: Human Shared Intentionality and Culture', in *Mind the Gap: Tracing the Origins of Human Universals*, ed. Peter M. Kappeler and Joan B. Silk (Berlin, 2010), pp. 331–50
— 'Understanding and Sharing Intentions: The Origins of Cultural Cognition', *Behavioral and Brain Sciences* 28.5 (2005), 675–91
Tooby, John, and Leda Cosmides, 'The Psychological Foundations of Culture', in *The Adapted Mind: Evolutional Psychology and the Generation of* Culture, ed. Jerome H. Barkow , Leda Cosmides and John Tooby (Oxford, 1992), pp. 19–136
Triandis, Harry C., 'The Self and Social Behaviour in Differing Cultural Contexts', *Psychological Review* 96.3 (1989), 506–20
Tribble, Evelyn, *Cognition in the Globe: Memory and Attention in Shakespeare's Theatre* (London, 2011)
— and Nicholas Keene, *Cognitive Ecologies and the History of Remembering: Religion, Education, and Memory in Early Modern England* (London, 2011)
Tsur, Reuven, 'Aspects of Cognitive Poetics', in *Cognitive Stylistics: Language and Cognition in Text Analysis*, ed. Elena Semino and Jonathan Culpeper (Amsterdam and Philadelphia, 2002), pp. 279–318
— *Poetic Rhythm: Structure and Performance* (Berne, 1998)
— 'Some Cognitive Foundations of "Cultural Programs"', *Poetics Today* 23.1 (2002), 64–89
— *Toward a Theory of Cognitive Poetics*, 2nd edn (Brighton and Portland, Or., 2008)
Turner, Mark, 'The Cognitive Study of Art, Language, and Literature', *Poetics Today* 23.1 (2002), 9–20
— *The Literary Mind: The Origins of Thought and Language* (New York and Oxford, 1996)
— *Reading Minds: The Study of English in the Age of Cognitive Science* (Princeton, 1991)
— and Giles Fauconnier, 'A Mechanism of Creativity', *Poetics Today* 20.3 (1999), 397–418
Tyler, Elizabeth, *Old English Poetics: The Aesthetics of the Familiar in Anglo-Saxon England* (York, 2006)

Ungerer, Friedrich, and Hans-Jörg Schmid, *An Introduction to Cognitive Linguistics* (London and New York, 1996)
Varela, F. J., E. Thompson and E. Rosch, *The Embodied Mind: Cognitive Science and Human Experience* (Cambridge, MA, 1991)
Vaughan, M. F., 'A Reconsidertation of "Unferth"', *Neuphilologische Mitteilungen* 77 (1976), 32–48
Velleman, J. D., *Self to Self: Selected Essays* (Cambridge, 2006)
Vermeule, Blakey, *Why Do We Care about Literary Characters?* (Baltimore, 2010)
Vickrey, John, 'Adam, Eve, and the *tacen* in *Genesis B*', *Philological Quarterly* 72 (1993), 1–14
—— '*Genesis* 549–51 and 623–25: Narrative Frame and Devilish Cunning', *Philological Quarterly* 76 (1997), 347–68
—— '*Genesis B* and the Anomalous Gnome', *Mediaevalia* 14 (1991 for 1988), 51–62
—— 'The Vision of Eve in *Genesis B*', *Speculum* 44 (1969), 86–102
Wan, Ching, K. Dach-Gruschow, S. No and Y. Hong, 'Self-Definitional Functions of Culture', in *Cultural Processes: A Social Psychological Perspective*, ed. Angela K. Leung, Chi-yue Chiu and Ying-yi Hong (Cambridge, 2011), pp. 111–35
Waterhouse, Ruth, 'Tone in Alfred's Version of Augustine's *Soliloquies*', in *Studies in Earlier Old English Prose*, ed. P. Szarmach (Albany, 1986), pp. 47–85
Werth, Paul, 'Extended Metaphor – A Text-World Account', *Language and Literature* 3.2 (1994), 79–103
—— *Text Worlds: Representing Textual Space in Discourse* (London, 1999)
Wertsch, James V., 'Collective Memory', in *Memory in Mind and Culture*, ed. P. Boyer and J. V. Wertsch (Cambridge, 2009), pp. 117–37
—— *Mind as Action* (Oxford, 1998)
—— *Voices of the Mind: A Sociocultural Approach to Mediated Action* (Cambridge, MA, 1991)
Wheelock, Jeremy I., 'The Word Made Flesh: "Engel Dryhtnes" in *The Dream of the Rood*', *English Language Notes* 37.3 (2000), 1–11
Whissell, Cynthia, 'The Flow of Emotion through Beowulf', *Psychological Reports* 99 (2006), 835–50
Whitelock, Dorothy, 'The Interpretation of "The Seafarer"', in *The Early Cultures of North-West Europe*, ed. Cyril Fox and Bruce Dickins (Cambridge, 1950), pp. 261–72
Wierzbicka, Anna, 'A Conceptual Basis for Cultural Psychology', *Ethos* 21.2 (1993), 205–31
—— *Emotions across Language and Cultures: Diversity and Universals* (Cambridge, 1999)
—— *Semantics, Culture and Cognition: Universal Human Concepts in Culture-Specific Configurations* (New York, 1992)
Wilcox, Miranda, 'Alfred's Epistemological Metaphors: *eagan modes* and *scip modes*', *Anglo-Saxon England* 35 (2006), 179–217
Williams, H. L., and M. A. Conway, 'Networks of Autobiographical Memories', in *Memory in Mind and Culture*, ed. P. Boyer and J. V. Wertsch (Cambridge, 2009), pp. 33–61
Williams, Raymond, *The Long Revolution* (Harmondsworth, 1965)
Woolf, Rosemary, 'The Fall of Man in *Genesis B* and the *Mystère d'Adam*', in *Studies in Old English in Honor of Arthur G. Brodeur*, ed. S. B. Greenfield (New York, 1963), pp. 187–99
Wordsworth, William, *Lyrical Ballads*, ed. R. L. Brett and A. R. Jones, 2nd edn (London, 1991)

Wouteres, Cas, 'The Civilizing of Emotions: Formalization and Informalization', in *Theorizing Emotions: Sociological Explorations and Applications*, ed. D. Hopkins, H. Kuzmics, H. Flam and J. Kleres (Frankfurt and New York, 2009), pp. 169–93

Wulff, Helena, ed., *The Emotions: A Cultural Reader* (Oxford and New York, 2007)

Young, Kay, *Imagining Minds: The Neuro-Aesthetic of Austen, Eliot, and Hardy* (Colombus, 2010)

—— and Jeffrey Saver, 'The Neurology of Narrative', *SubStance* 30 (2001), 72–84

Zlatev, Jordan, T. P. Racine, C. Sinha and E. Itkonen, eds, *The Shared Mind: Perspectives on Intersubjectivity* (Amsterdam, 2008)

Znojemská, Helena, 'Medieval English Soul and Body Literature', *Litteraria Pragensia: Studies in Literature and Culture* 17.34 (2007), 20–39

Zunshine, Lisa, 'Fiction and Theory of Mind: An Exchange', *Philosophy and Literature* 31.1 (2007), 189–96

—— 'Lying Bodies of the Enlightenment: Theory of Mind and Cultural Historicism', in *Introduction to Cognitive Cultural Studies*, ed. Lisa Zunshine (Baltimore, 2010), pp. 115–33

—— 'Richardson's *Clarissa* and a Theory of Mind', in *The Work of Fiction: Cognition, Culture and Complexity*, ed. Alan Richardson and Ellen Spolsky (Aldershot and Burlington, 2004), pp. 127–46

—— 'Theory of Mind and Experimental Representations of Fictional Consciousness', *Narrative* 11 (2003), 270–91

—— 'Theory of Mind and Fiction of Embodied Transparency', *Narrative* 16.1 (2008), 65–92

—— 'Why Jane Austen was Different, and Why We May Need Cognitive Science to See it', *Style* 41.3 (2007), 275–99

—— *Why We Read Fiction: Theory of Mind and the Novel* (Columbus, 2006)

—— ed., *Introduction to Cognitive Cultural Studies* (Baltimore, 2010)

Index

ANGLO-SAXON STUDIES

Volume 1: The Dramatic Liturgy of Anglo-Saxon England,
M. Bradford Bedingfield

Volume 2: The Art of the Anglo-Saxon Goldsmith: Fine Metalwork in Anglo-Saxon England: its Practice and Practitioners,
Elizabeth Coatsworth and Michael Pinder

Volume 3: The Ruler Portraits of Anglo-Saxon England,
Catherine E. Karkov

Volume 4: Dying and Death in Later Anglo-Saxon England,
Victoria Thompson

Volume 5: Landscapes of Monastic Foundation:
The Establishment of Religious Houses in East Anglia, c. 650–1200,
Tim Pestell

Volume 6: Pastoral Care in Late Anglo-Saxon England,
edited by Francesca Tinti

Volume 7: Episcopal Culture in Late Anglo-Saxon England,
Mary Frances Giandrea

Volume 8: Elves in Anglo-Saxon England: Matters of Belief, Health, Gender and Identity, *Alaric Hall*

Volume 9: Feasting the Dead: Food and Drink in Anglo-Saxon Burial Rituals,
Christina Lee

Volume 10: Anglo-Saxon Button Brooches:
Typology, Genealogy, Chronology,
Seiichi Suzuki

Volume 11: Wasperton: A Roman, British and Anglo-Saxon Community in Central England,
edited by Martin Carver with Catherine Hills and Jonathan Scheschkewitz

Volume 12: A Companion to Bede,
George Hardin Brown

Volume 13: Trees in Anglo-Saxon England: Literature, Law and Landscape,
Della Hooke

Volume 14: The Homiletic Writings of Archbishop Wulfstan,
Joyce Tally Lionarons

Volume 15: The Archaeology of the East Anglian Conversion,
Richard Hoggett

Volume 16: The Old English Version of Bede's *Historia Ecclesiastica*,
Sharon M. Rowley

Volume 17: Writing Power in Anglo-Saxon England:
Texts, Hierarchies, Economies,
Catherine A. M. Clarke

Printed and bound by CPI Group (UK) Ltd, Croydon, CR0 4YY

13/09/2026

14960995-0001